The Forager's Feast

THE COUNTRYMAN PRESS
A division of W. W. Norton & Company
Independent Publishers Since 1923

The Forager's Feast

How to Identify, Gather,
and Prepare Wild Edibles

LEDA
MEREDITH

COUNTRY MAN KNOW HOW

For information about permission to reproduce selections from
this book, write to Permissions, The Countryman Press,
500 Fifth Avenue, New York, NY 10110

For information about special discounts for bulk purchases,
please contact W. W. Norton Special Sales at
specialsales@wwnorton.com or 800-233-4830

Library of Congress Cataloging-in-Publication Data

Names: Meredith, Leda, author.
Title: The forager's feast : how to identify, gather, and prepare
wild edibles / Leda Meredith.
Description: Woodstock, VT : The Countryman Press,
a division of W. W. Norton & Company, [2016] | Series:
Countryman know-how | Includes index.
Identifiers: LCCN 2015040712 | ISBN 9781581573060 (pbk.)
Subjects: LCSH: Cooking (Wild foods) | Cooking (Natural foods)
| Wild plants, Edible—North America. | Forage plants—North
America. | LCGFT: Cookbooks.
Classification: LCC TX741 .M47 2016 | DDC 641.3/02—dc23
LC record available at http://lccn.loc.gov/2015040712

THE COUNTRYMAN PRESS
www.countrymanpress.com

A division of W. W. Norton & Company
500 Fifth Avenue, New York, NY 10110
www.wwnorton.com

10 9 8 7 6 5 4 3 2 1

DEDICATION

To everyone who ate something that grew wild today,
something you harvested with care, prepared
creatively, and shared with love.

CONTENTS

Foraging in the 21st Century

Odds are that you don't have to forage for edible wild foods in order to survive. In the 21st century, more than half of humanity lives in cities or suburbs. The cultural knowledge of which plants, algae, and fungi are safe to eat has been relegated to a hobby. Or has it?

I've witnessed an unmistakable surge of interest in foraging over the past few years. Participation in foraging tours is at an all-time high, not just for mine, but also for colleagues with whom I've discussed this upsurge. Book publishers (including this book's publisher) are seeking foragers who can write, restaurants are making a big deal about the wild ingredients in their menus, and tens of thousands of foragers and wannabe foragers have joined social media groups devoted to the topic.

Why the surge in interest? I believe it is because foraging encompasses an entire crossroads of immediate concerns.

Let's start with the obvious: it's free food.

Economic fluctuations in the 21st century have been epic so far, and with every downturn there has been an upswing of interest in self-sufficiency. Knowing what you can pick for free and serve for dinner tonight is the ultimate in DIY skills.

Another branch of the crossroads is foodie interest in tasty ingredients that might not be available for sale anywhere. When was the last time you picked up a pint of juneberries at the supermarket, or paid cash for cattail laterals? You didn't, unless you paid for a dish at one of the many restaurants now employing professional foragers to bring them unusual ingredients, or unless you went out and foraged your own.

Foraging is also topical for anyone concerned about environmental footprints (which should be everyone nowadays). Done correctly, foraging actually improves the balance of local ecosystems by reducing the number of invasive species and encouraging—through correct harvesting methods—slow-growing and endangered species.

As far as eating local and organic, buzzwords that reflect widespread concern for both environmental issues and personal health, nothing is more local than the edible weed growing in your backyard. And nothing is more "organic" than a wild plant thriving without fertilizers, herbicides, or other human intervention (assuming it was collected from a location free of such chemical products).

Wild foods are often nutritionally superior to their cultivated counterparts, making physical health yet another reason to forage. Besides the fact that you'll be outside getting some exercise (never a bad thing), you'll be eating foods like lamb's quarters, which has three times the calcium and more than double the vitamin C of its genetic cousin, spinach.

There are other, less tangible but equally valid reasons for foraging. It sounds hippy-dippy, but a direct connection to nature and your immediate environment is something you can experience quite tangibly through foraging. The benefits of that connection include a state of well-being that is both peaceful and exuberant. If you don't know what I mean, think about picking blackberries on a warm summer morning and trying not to eat them all before you get home . . . compared with purchasing an expensive pint of blackberries in a plastic container from the refrigerated section of the supermarket.

For all of these reasons and more, I am delighted that you are enthusiastic enough about foraging to have picked up this book!

The Forager's Feast includes recipes that range from classics such as garlic mustard pesto and wood sorrel soup, to inventions that may pique the interest of even the most jaded gourmet forager such as popped amaranth crackers, pineappleweed cordial, and wild kimchi powder.

In order to make room for all the recipes, I had to limit the number of edible wild plants I profile in the book. Included here are only plants that are widely distributed (in most cases more than one continent) and easy to identify. In other words, if you live on planet Earth I guarantee that some of these plants are growing wild near you and that you will be able to recognize them with 100 percent certainty with the help of this book.

Essential to 21st-century foraging is an understanding of how your harvesting impacts the natural environment. Each plant account offers tips on how to harvest that ingredient while ensuring there will be plenty more to come back for next year. The ecosystem will actually be better off because you foraged there.

How to Use This Book

The Forager's Feast has everything you need to help you easily find, safely identify, sustainably harvest, and deliciously prepare dozens of the most widespread edible wild plants. Here are some tips to help you get the most out of your foraging adventures.

FINDING THE WILD EDIBLES

Start with the Which Wild Edible Where and When section. This will make you a much more successful forager than you would be if you wandered out with no clue as to which plants you can find in that particular location during that exact season. For example, if you set out in spring looking for ripe blackberries in a spot fully shaded by evergreen trees, you are doomed to frustration: blackberries are a summer fruit requiring at least partial sunlight. You might be able to identify the blackberry canes without their fruit, and that will make you a super-efficient forager come summer, when you can beeline for the blackberry patch. But you'd be going home empty-handed on that early spring day.

IDENTIFYING THE WILD EDIBLES

If you checked the seasonal guide when you headed out in April, you saw that fresh spruce tree tips were in season in spring and you'd have been able to verify whether those evergreen trees shading you out were, in fact, spruce trees. Bingo! Now, instead of wasting time hunting for out-of-season berries, you're going home with a bag full of one of nature's

tastiest seasonings, available for only a few weeks early in the year.

But the most important thing is to double-check the characteristics of the plant you're looking at with the description in this book. Yes, the photos are helpful, but matching the description is more important.

FORAGER'S RULE #1

Always be 100 percent certain of your identification before consuming any wild edibles, seaweed, or mushroom.

There's a saying that there are old foragers and there are bold foragers, but there are no old, bold foragers. Take that to heart. Although I want to encourage you to get out there and forage, I also want you to be safe. What does that mean? It means that *all* of the identifying characteristics of the wild edible must match the description given in that wild edible's entry. But don't worry: identifying wild edibles isn't actually that hard.

Can you tell the difference between the bunches of parsley and cilantro (coriander) next to each other at the market? Well, that is plant identification. And I bet you used your nose as well as your eyes to tell the difference between them, right? Good! Get your senses of sight, smell, and touch involved in your plant ID and save the sense of taste for *after* you've clinched the ID, okay?

Also, remember that location is an important part

of plant and mushroom identification. If you tell me that you found a full-sun plant growing in deep forest shade, I'm going to be pretty certain that you have misidentified that plant. For that reason, the Which Wild Edible Where and When section is very location specific, as is the Find section in each wild edible's individual profile.

I have tried to keep the plant descriptions crystal clear without resorting to too much technical jargon, but occasionally it was unavoidable. Leaf arrangements, for example, are an essential part of plant identification. Here's a quick guide.

LEAF ARRANGEMENTS

It is not just the shape of the leaves that matters when identifying a plant. The way the leaves are arranged on the stem is also an important identification characteristic. Here are the main types of leaf arrangements:

Opposite
The leaves join the stems in pairs.

Alternate
The leaves join the stems singly, with each leaf farther along the stem than the one before it.

Whorled
Three or more leaves attach at the same point on the stem, often forming a circle of leaves around that stem node.

Rosette

Rather than growing along a stem, rosette leaves form a circle, with the bases of the leaves all going into the ground at a central point where they attach to the root (think dandelion).

SAFELY AND SUSTAINABLY HARVEST

Some plants multiply so voraciously that they are considered invasive weeds. Most often these species were introduced from other continents, and they can crowd out slower-growing native species. Fortunately for foragers, many invasive plants are delectable wild edibles. You can harvest these plants freely without any worry about sustainability.

On the other hand, what about those slower-growing, noninvasive plants? These should be harvested only where they are abundant, and you should always leave the majority of the patch alone so that it can regenerate. Rather than picking everything from one plant or clump in front of you, think

of yourself as a grazing animal and collect a little here, a little there, leaving plenty behind. In some cases there are specific techniques for harvesting certain plants that ensure the plant population isn't threatened by your collection methods.

I'll let you know whether a wild edible is invasive and what the best sustainable harvesting technique is in each species' profile section.

The next thing to consider is whether the location is polluted or relatively toxin-free. Here's your checklist:

· Have herbicides or pesticides been sprayed in the area? Industrial farms, most suburban front yards, railroad right-of-ways, and almost all golf courses are often heavily sprayed. City parks are also sometimes sprayed.

· Are the plants growing close to a heavily trafficked road? Pick at least 50 feet away from such roads, more if you are downhill from traffic.

· When collecting plants that grow near water, is the water clean enough to drink safely? If not, you might still be able to eat the plant once it is thoroughly cooked, but you should not taste it raw.

Okay, let's assume all of the identification characteristics of the plant you're looking at match the description in this book or another reliable field guide. You're 100 percent sure which plant you've found, and it is edible. The spot where it is growing is relatively pollution-free. (I say "relatively" because in the 21st century it is very difficult to find any location, even in the wilderness, that is 100 percent without traces of pollution.)

Time to eat, right? Not just yet. There are three more things to make sure of before you can chow down:

- Which part of the plant is edible? It is not uncommon for a plant to have one part that is edible and another that is toxic. Elderberry, for instance, has delicious, safe-to-eat flowers and fruit, but its leaves are poisonous.

- At what stage is the plant edible? At what stage is it palatable? Dandelion greens in early spring are so good you'll want a second helping, but pick them in summer and you'll probably spit them out because they are very bitter that late in the year.

- How is the plant prepared? While many wild edibles are good both raw and cooked, some are only safe to eat once they've been cooked.

I answer these questions for you in each plant's individual description.

Collecting and Processing Techniques

LEAFY GREENS

Because so many leafy greens are unpleasantly bitter during the warm months of the year, it's always worth taking a small bite to find out if it is worth harvesting more at that time. If the taste is pleasant, the next thing to remember is that cooked leafy greens reduce drastically in size: 10 cups of raw leafy greens will reduce down to about 1 cup of cooked greens on average. So harvest *a lot* if you plan to cook them rather than eat them raw.

The next page offers a good basic recipe for cooked wild leafy greens.

Universal Leafy Greens Recipe

Serves 4

This is my version of everyone's go-to way to cook leafy greens. It is how my grandma and great-grandma made them when I was a kid.

INGREDIENTS

1½ pounds leafy greens, weighed after removing any tough stems

2 tablespoons extra-virgin olive oil

2 garlic cloves, peeled and minced or pressed

1 lemon, cut into wedges

Salt and freshly ground black pepper to taste

INSTRUCTIONS

1. Rinse the greens clean and put them into a large skillet with the rinsing water still clinging to them. Cook them over medium heat, stirring often, until wilted. Add more water if the leaves start sticking to the pan.

2. Add the oil and the garlic and cook, stirring, for 2 minutes more.

3. Serve hot or at room temperature with the lemon wedges and salt and pepper so that you and your guests can adjust the seasoning to taste.

STALKS AND SHOOTS

You know how we are taught to tell the tender part of an asparagus shoot from the tough part by holding both ends and bending until the spear naturally snaps at the dividing line between the two? Well, that dividing line between tender and tough exists on wild stalks and shoots as well.

Start at the base and bend the stalk gently, then do the same a little farther up, then a few inches farther up, until it snaps off cleanly. If you get all the way to the tip and it never snaps off easily, the stem is probably too tough to use.

An exception to this is the immature flower stalk of a second-year burdock plant (my favorite edible part of that plant). Its stalk is tough enough to require a knife to remove it from the plant, so instead of the bending process above, use the "rule of thumb" below.

The "rule of thumb" states that if you can easily pierce shoots or stalks with your thumbnail, they will be tender enough to eat. If not, they are probably going to be too tough and stringy to be good.

Note that sometimes even stems and shoots that are slightly too tough to pass the bend or thumbnail tests are still tender enough to eat if you peel them before you cook them.

WINNOWING WILD SEEDS

Nobody wants a mouthful of chaff along with their seed harvest. Winnow the chaff out of lamb's quarters, amaranth, and other wild seeds by first drying the seedheads for several days or weeks in paper

or cloth bags. Take a big bowl of dried seedheads outside on a breezy day (or set up in front of a fan at its lowest speed and adjust your distance.) Keep in mind that the chaff will blow everywhere, so this is still best done outside.

Rub the seedheads between your hands to release the seeds. Scoop up handfuls of the seed and chaff and slowly pour it back into the bowl from about a foot above the bowl. The chaff will blow away while the heavier seeds drop back into the bowl. Don't worry about separating 100 percent of the chaff from the tiny seeds, but do try to get most of it out.

Which Wild Edible Where and When

This section will make you a more efficient forager. Use this information to pinpoint what you can expect to find at any given time of year in specific types of locations.

Be aware, though, that the timing of wild harvests varies depending on your region. For example, in southern regions the redbuds may finish blooming by April, whereas farther north the same species will be just starting to bloom then.

Another variable is the unique weather of any particular year. If it was an especially cold winter and the temperature warmed up a few weeks later than usual, expect many of the plants to come into season later as well.

EARLY SPRING

Meadows, Fields, Disturbed Soils, Sunny Areas, and at the Edges of Sunny Areas

Burdock root; Chickweed; Chicory leaves and roots; Dandelion leaves, crowns, and roots; Dock leaves; Garlic Mustard leaves; Henbit; Jerusalem Artichoke; Japanese Knotweed shoots; Mint; Morel mushroom; Mugwort; Mustard greens; Nettles; Plantain leaves; Redbud blossoms; Shepherd's Purse leaves; Violet leaves and flowers; Wild Garlic

Forests and Partially Shaded Places

Black Birch or Yellow Birch inner bark and twigs; Chickweed; Garlic Mustard leaves and roots;

Clover

Japanese Knotweed shoots; Mint; Morel mushroom; Nettles; Oyster mushroom; Redbud blossoms; Spruce tips; Violet leaves and flowers; Wild Garlic

Seashore and Coastal Areas

Dulse; Sea Lettuce

Wetlands, River Banks, Lakesides, Bogs

Black Birch or Yellow Birch inner bark and twigs; Cattail shoots and lateral rhizomes; Mint

MIDSPRING TO LATE SPRING

Meadows, Fields, Disturbed Soils, Sunny Areas, and at the Edges of Sunny Areas

Amaranth leaves; Asiatic Dayflower; Burdock roots and immature flower stalks; Cattail shoots and lateral rhizomes; Chickweed; Chicory roots; Clover blossoms; Dandelion flowers and roots; Dock flower stalks; Grape leaves; Juneberries; Lamb's Quarters leaves; Mallow; Mint; Morel mushroom; Mugwort; Mulberry; Nettles; Peppergrass; Pineappleweed; Plantain young leaves; Purslane; Quickweed; Sheep Sorrel; Shepherd's Purse shoots; Shiso; Sow Thistle leaves and shoots; Wild Grape leaves; Wild Lettuce; Wood Sorrel

Forests and Partially Shaded Places

Black Birch or Yellow Birch inner bark and twigs; Chickweed; Garlic Mustard leaves, shoots, flowers, and immature seedpods; Juneberries; Morel mushroom; Oyster mushroom; Spruce tips; Violet young leaves; Wood Sorrel

Seashore and Coastal Areas

Dock leaves; Dulse; Sea Lettuce; Sheep Sorrel

Wetlands, River Banks, Lakesides, Bogs

Black Birch or Yellow Birch inner bark and twigs; Cattail shoots and rhizomes; Mint

SUMMER

Meadows, Fields, Disturbed Soils, Sunny Areas, and at the Edges of Sunny Areas

Amaranth leaves; Apple; Asiatic Dayflower; Blackberry; Chickweed; Chicory roots; Clover; Dandelion roots; Dock seeds; Grape leaves and fruit; Hawthorn; Lamb's Quarters leaves; Mallow; Mint; Mugwort; Mulberry; Mustard seeds; Peppergrass; Pineappleweed; Plantain young leaves and seeds; Purslane; Quickweed; Rose leaves and flowers; Sheep Sorrel; Shepherd's Purse; Shiso; Sumac; Wood Sorrel

Forests and Partially Shaded Places

Asiatic Dayflower; Blackberry; Chicken of the Woods mushroom; Chickweed; Garlic Mustard seeds; Grape leaves and fruit; Hawthorn; Maitake/Hen of the Woods mushroom; Mint; Oyster mushroom; Wood Sorrel

Blackberries

Wild grapes

Seashore and Coastal Areas
Dock seeds; Dulse; Sea Lettuce; Sheep Sorrel

Wetlands, River Banks, Lakesides, Bogs
Black Birch or Yellow Birch inner bark and twigs; Cattail pollen and immature seedheads; mint

AUTUMN
Meadows, Fields, Disturbed Soils, Sunny Areas, and at the Edges of Sunny Areas
Amaranth seeds; Apple and Crabapple; Asiatic Dayflower; Chickweed; Chicory roots and leaves; Dandelion roots; Dock leaves and seeds; Hawthorn; Henbit; Jerusalem Artichoke; Lamb's Quarters

seeds; Mallow; Mint; Oak acorns; Quickweed; Rose hips; Wild Garlic; Wild Grapes

Forests and Partially Shaded Places
Black Birch or Yellow Birch inner bark and twigs; Chicken of the Woods mushroom; Chickweed; Wild Garlic; Garlic Mustard roots and leaves; Hawthorn; Maitake/Hen of the Woods mushroom; Oyster mushroom

Seashore and Coastal Areas
Dock leaves and seeds; Dulse; Rose hips; Sea Lettuce

Wetlands, River Banks, Lakesides, Bogs
Black Birch or Yellow Birch inner bark and twigs; Cattail; Mint

WINTER
Meadows, Fields, Disturbed Soils, Sunny Areas, and at the Edges of Sunny Areas
Chickweed; Crabapple; Henbit; Jerusalem Artichoke; Rose hips; Sumac; Wild Garlic

Forests and Partially Shaded Places
Birch; Chickweed; Oyster mushroom; Wild Garlic

Seashore and Coastal Areas
Dock seeds; Rose hips; Sea Lettuce

Wetlands, River Banks, Lakesides, Bogs
Birch; Cattail rhizomes

Wild garlic

The Wild Edibles

All of the following edible wild species are easy to find and identify across many regions of the world. I promise you will be able to find some of them growing near you! Where I have given the Latin genus name but not listed a particular species, it means that there is more than one edible species within that genus that has the characteristics listed in the description.

This amaranth plant's seeds are just starting to ripen.

Amaranth
AMARANTHUS SPECIES
Once a staple food in Central and South America, amaranth offers foragers both a leafy green vegetable and a high-protein seed that can be cooked like a whole grain.

Find
Amaranth doesn't show up until midspring, when nighttime temperatures are consistently above freezing. It likes full sun and is a common farm and garden weed. It is equally happy growing in cracks on city streets as it is in the country.

Identify
The size of different plants within the *Amaranthus* genus varies from ground-hugging, mat-forming species to some that grow as tall as 7 feet. All of them have edible leaves and seeds.

The oval leaves have smooth edges except for a single notch at the tip of the leaf in some *Amaranthus* species. The leaves attach to the stalks in an alternate pattern. The veins are quite prominent.

Brush the soil away from the top of an amaranth plant's root and you may see that it is a pink or red color. (Redroot is one of several names for the most widely distributed amaranth of all, Common Amaranth, *Amaranthus retroflexus*). Other amaranth species may or may not have reddish roots. Sometimes their stems and the undersides of their leaves will have this reddish coloration as well.

The fuzzy clusters of flowers and eventually seeds grow in the leaf axils and/or the tops of the plants. They start out looking like thick green clusters and then turn a buff or brown color as the seeds mature. Wild amaranth seeds are tiny and usually a shiny black or brown, unlike the light-colored amaranth seeds you can buy at health food stores.

Sustainably Harvest
All amaranth species self-seed prolifically and are generally considered invasive. In the 1500s, Cortés realized how important the amaranth crops were to the Aztecs and destroyed many of their cultivated amaranth fields. Despite this, amaranth has since spread to four other continents. (Clearly, amaranth had the last laugh.) You do not have to worry about overharvesting this plant.

Collect the leaves anytime from when they first appear in spring until the plants start to flower in late summer.

Harvest the seeds once they are fully matured. At that stage, the fuzzy clusters will no longer be green, and the seeds will fall out easily if you lightly crush the seedheads over a container. I like to snip the ripe seedheads off directly into a paper or cloth bag and let them dry for about a week. After that, use the winnowing method in the Techniques section (page 16) to remove the chaff from the seeds.

Just one warning: Avoid collecting seedheads from Spiny Amaranth, *Amaranthus spinosus*. The sharply pointed awns are impossible to completely remove from the winnowed product. They will stick inside one's mouth and throat like tiny, sharp splinters.

Eat

Amaranth leaves are edible raw, but in my opinion they are much better cooked. They are mild flavored and good in any recipe that would work with spinach or kale.

Although technically a seed rather than a grain (grains come from plants in the Fabaceae and Poaceae families, and amaranth is not one of those), amaranth seeds are usually treated like a grain in the kitchen. If you boil them, use more liquid than you would with rice or other grains: 2½ to 3 cups of liquid per cup of amaranth seeds is a good ratio. As a cooked "grain," I usually prefer amaranth in combination with another whole grain.

Amaranth seeds also can be popped like popcorn. The popped seeds can then be enjoyed as a breakfast cereal, crunchy salad topping, or in these crackers.

Popped Amaranth Seeds for Popped Amaranth Crackers

These versatile crackers have a mild, nutty flavor that goes well with just about any topping. They get their lightness and crispness from popped amaranth, so let's get popping.

HOW TO POP AMARANTH SEEDS

You don't need to use oil to pop amaranth's tiny seeds, but you do need to get familiar with your stove's idiosyncrasies. The heat needs to be high enough to pop the seeds quickly, but not so high that they burn before they pop. Be prepared to burn a batch or two until you find the perfect setting on your stove. Once you've got that figured out, popping amaranth is easy. It also helps to use a heavy-bottomed pot rather than a flimsy one.

INSTRUCTIONS

1. Start by heating a large pot. It's important that it be hot before you add the amaranth. A drop of water should sizzle and evaporate within a second or two.

2. Add just 1 to 2 tablespoons of amaranth seeds per batch. Even though you're using a large pot, don't try adding more than this much amaranth or it will burn. Shake the pot to spread the seeds out in a single layer. Cover the pot. (It's helpful if the pot has a glass lid so that you can see what's going on during the popping.)

3. The seeds should start popping within 10 seconds if the pot is correctly preheated. If they burn without ever popping, your heat is too high; try again over lower heat. Shake the pot occasionally while they pop.

4. As soon as the popping slows down, remove the pot from the heat. Not all of the seeds will pop— that's okay. The ones that don't will still be nicely toasted but not burnt.

5. Empty the popped amaranth into a bowl. Return the pot to the stove and let it reheat for half a minute before adding the next tablespoon of amaranth seeds.

6. If you plan on storing the popped seeds (they make a great addition to breakfast cereal), let them cool completely before transferring to an airtight container.

Popped Amaranth Crackers

Makes about 20 crackers

INGREDIENTS

¾ cup amaranth flour (you can make this easily by grinding amaranth seeds in an electric coffee grinder)

⅓ cup Popped Amaranth Seeds

½ teaspoon salt

¼ teaspoon garlic powder (optional)

2 tablespoons coconut oil, melted butter, or extra-virgin olive oil (which fat you choose completely changes the flavor of the crackers, but they're all good)

¼ cup water

INSTRUCTIONS

1. Preheat the oven to 400°F. Put a baking sheet–size piece of parchment paper on your work surface.

2. In a large bowl, whisk together the amaranth flour, popped seeds, salt, and garlic powder. Add the oil and work it in with your clean fingers until the mixture is something like the texture of coarse cornmeal.

3. Add the water a little at a time, stopping when the dough is still crumbly but holds together when you squeeze a pinch of it.

4. Scoop the dough out onto the parchment paper. You can either press the dough out with your hands or use a rolling pin. If you use a rolling pin, you will need to dust the dough occasionally with a little flour so that it doesn't stick (you can use more amaranth flour or all-purpose flour). Keep going until the dough is spread out and between ⅛- and ¼-inch thick.

5. Carefully transfer the parchment paper with the dough on it to a baking sheet. Using a table knife, score the dough in a tic-tac-toe pattern so that it will break into squares or rectangles after it is baked.

6. Bake for approximately 10 minutes, until the edges are browned but not burnt. If you rolled the dough out on the thick side, it may take a little longer.

7. Take the baking sheet out of the oven and cool completely before breaking into individual crackers.

Eggs Amaranth

Makes 2 to 4 servings (depending on whether you give each person one egg or two)

Eggs Florentine shows up frequently on restaurant menus. It is always some version of cooked spinach plus eggs. This is the same idea, except it uses cooked amaranth leaves instead of spinach, and I've given a nod to amaranth's Central and South American origins with the seasoning.

INGREDIENTS

2 ounces bacon

1 small onion, finely chopped

2 cloves garlic, minced or pressed

½ pound fresh amaranth leaves

¾ cup heavy cream

¾ teaspoon ground cumin

¼ teaspoon freshly ground nutmeg

¼ teaspoon ground cayenne (optional)

½ cup grated Romano cheese

Salt and pepper to taste

2 large tomatoes

4 eggs

Lemon juice or vinegar if hand poaching (see page 30)

Eggs amaranth is a company-worthy brunch dish.

INSTRUCTIONS

1. Cut the bacon into small bits or strips and render it in a skillet over low heat. When most of the fat has been released and the bacon is getting crisp (but not burnt), transfer the cooked bacon to a dish. If

continued

29

there's a lot of bacon fat, drain off most of it, but leave enough to coat the bottom of the skillet.

2. Cook the onion and garlic in the residual bacon fat over medium-low heat, stirring frequently until the onion softens and becomes translucent.

3. While the *Alliums* (onion and garlic) are doing their thing, bring a large pot of water to a boil. Add the amaranth leaves and cook over high heat for 5 minutes. Drain in a colander, then run under cold water to stop any residual cooking. Squeeze out as much liquid as you can. Add the cooked amaranth to the onion and garlic. Add the cream, cumin, nutmeg, and cayenne and cook, stirring, until the sauce thickens. Remove from heat. Stir in the grated cheese and salt and pepper to taste.

4. Preheat oven or toaster oven on the broil setting. Line a baking sheet with parchment paper. Slice the tomatoes ¾-inch thick. Compost the stem and tip ends. Arrange the remaining slices on the sheet and broil for 8 minutes. You can keep the tomatoes warm in a low oven (250°F) for as long as it takes to complete the next step.

5. While the tomatoes are in the broiler, poach the eggs.

6. Assemble the Eggs Amaranth: Place two of the tomato slices on each plate. Spoon some of the creamed, spiced amaranth greens over each tomato. Place the poached eggs on top of the amaranth. Sprinkle the bacon over the top. Serve immediately.

POACHING EGGS

If you have an egg poacher, this is easy: Put about an inch of water in the pan. Lightly butter the egg cups. Crack one egg into each buttered cup and place the cup holder on top of the water in the pan. Cover and cook over high heat until the whites are opaque and mostly set, but the yolks still jiggle when the pan is shaken. Immediately remove the lid and remove from the heat. There will be some further cooking from the residual heat, so err on the side of underdone when you take them out.

To poach the eggs without an egg poacher, put about an inch of water into a medium-large skillet (a 12-inch diameter is perfect). Add 1 teaspoon of salt and 2 teaspoons of lemon juice or white or cider vinegar to the water. While the water is coming to a simmer, crack the eggs into individual ramekins or cups. When the water is ready, carefully pour each egg into the water. Turn off the heat, cover the pan, and let the eggs poach in the hot water for 5 minutes. Remove them carefully with a slotted spoon.

Tip: You can keep poached eggs in a bowl of ice water and refrigerate them for up to 8 hours. Reheat them in warm water just before serving.

Other plants you can use in this recipe: lamb's quarters, Asiatic dayflower, quickweed, nettles, and any mildly flavored, cooked leafy green.

Apples and Crabapples
MALUS SPECIES

Wild apples and crabapples are often more tart than cultivated apples, but they make fantastic cider and vinegar. They are high in pectin and also make great preserves, including apple jelly.

Find

Look for wild apples and crabapples where they were once intentionally planted and their fruit forgotten (the case with many park apples and landscaped crabapples), and where they have grown wild from scattered fruit. You'll also find them on sunny, grassy slopes and roadsides near orchards, both active and abandoned. I find that most property owners are happy to let foragers gather fruit that would otherwise go to waste, assuming you ask permission first. Most likely, you will be welcome to pick as long as you don't break a branch or leave a mess. And you may be invited back next year if you offer to pick up the apples littering the ground beneath the tree.

Identify

The trees can get to be 30 feet tall but are usually much smaller than that, often with bent and gnarled trunks. Both apples and crabapples have craggy, gray bark and alternate, unlobed, lightly toothed and pointed leaves with hairs that make them feel like felt on their undersides. All apples have five-petaled pink or white flowers and five seeds per fruit that form a pentacle pattern.

Crabapples are usually not more than ¾ inch in diameter and hang in long-stemmed clusters.

Sustainably Harvest

As with other tree fruit, you are not harming the parent plant by picking the fruit.

You'll know the fruit of an apple tree is ready when the first apples start voluntarily dropping to the ground. Look down to see when this has begun (anytime from midsummer through fall), then up to pick the apples that are still on the branches.

With crabapples it's harder to tell because they often stay on the tree right through winter. First wait for them to change completely from green to gold or red. Then taste one: Unripe crabapples are more astringent than sour, with a bitter edge that will make your mouth pucker. Ripe crabapples are still super sour, but without that astringent edge.

Eat

Some wild apples are sweet enough to eat raw, and the only way you'll know if you've got one of those is to taste test. Sweet wild apples can be used in pies and for sauce just like cultivated apples. They also make good fruit leather and are an excellent snack when dried.

Most wild apples, and every crabapple I've ever encountered, are too sour to be really good raw. But the sour apples, much more than the sweet ones, are super high in the pectin that makes jellies and jams gel. You can use them to boost the gelling power of low-pectin fruits and thus avoid buying store pectin.

Sour apples are also the best for making cider, and wild cider has the potential to become the best apple cider vinegar. You don't necessarily need a cider press for those projects: for small household batches, an electric juicer will do.

Easy Apple Cider

Yes, you can do this with any raw apple juice. But ripe-yet-tart wild apples produce a cider superior to any made from purely sugary cultivated apples.

INGREDIENTS

Wild apples

INSTRUCTIONS

1. Wash the apples and chop them into chunks.

2. Run them through an electric juicer.

3. Put the juice into a clean, wide-mouthed vessel like a bucket or a crock. Cover with a dish towel to keep out fruit flies. Leave at room temperature, stirring vigorously several times a day (the stirring is important).

4. After a day or two or three, the juice should start to froth up every time you stir it and begin to smell alcoholic. Transfer it to beer bottles or jugs, cap tightly, and store in your refrigerator or another cold place. Let age for 3 weeks before drinking. That's it.

If you let your cider sit out at room temperature for extra days in the wide-mouthed vessel, it will start to convert from a lightly alcoholic liquid to vinegar. This can be a good thing. Just leave it out until it is as sour as you'd like it, then bottle it. Note that a "mother" of vinegar will form. The mother looks like a gelatinous blob, but it is not dangerous. In fact, it's a sign that the vinegar bacteria are doing a brilliant job of converting alcohol into vinegar.

Crabapple Jelly

Makes 2 to 3 half-pint jars

When I was a child there was an old lady who used to give me jars of her golden, translucent crabapple jelly each year. The first time I harvested crabapples I knew that I wanted to try to re-create her wonderful gift.

INGREDIENTS

1½ pounds crabapples

Water

1 to 1½ cups sugar

2 tablespoons lemon juice

INSTRUCTIONS

1. Chop the crabapples and put them into a large pot. Add just enough water to cover them. Bring to a boil over high heat and cook until the water has evaporated by half. The crabapples should be getting mushy at this point.

2. Strain through a jelly bag or any cloth bag (a piece of a clean, old sheet works). Do not squeeze or your jelly will be cloudy. Let the liquid drain with no help from you for several hours. You'll need to suspend the bag over a pot: one way to do this is to tie it to the faucet of your kitchen sink with a pot underneath.

3. Measure the liquid and put it into a large pot. Add ½ cup sugar and 1 tablespoon lemon juice for each cup of crabapple extraction.

4. Bring the ingredients to a boil over high heat, stirring often. When the mixture reaches the gel point (see Resources for websites and books that will help you determine when that is), remove the pot from the heat. Pour the hot jelly into sterilized canning jars, leaving ½ inch of headspace between the surface of the jelly and the rims of the jars. Wipe the rims clean and screw on the canning lids.

5. Process in a boiling water bath for 5 minutes (adjust the time if you live at a high altitude—again, check Resources for guidance if you're unfamiliar with canning).

Asiatic Dayflower
COMMELINA COMMUNIS AND OTHER *COMMELINA* SPECIES

Never bitter and with a hint of pealike sweetness, this common garden weed offers four edible parts that are in season for several months.

Find

Dayflowers prefer partial shade and moist soil. They often grow at the edges of paths and in gardens, parks, and other habitats where humans have been busy. Asiatic dayflower doesn't appear until the warm temperatures of late spring, but it is in season all the way through late summer and sometimes early fall.

Identify

The first thing you'll probably notice about Asiatic dayflower is its flowers, which look like tiny irises. With *Commelina communis*, the two upper petals are blue to purplish-blue, and the third lower petal is white. In several other *Commelina* species all three petals are blue.

Commelina leaves are 1 to 4 inches long, lance-shaped, with smooth margins. They can be anywhere from ½ to 1½ inches wide, and are sometimes slightly hairy, especially at the stem-clasping bases of the leaves.

The juicy, almost succulent stems are swollen at the nodes where the leaves join the stems. Individual stems can grow to be 2½ feet long, but because of the plant's sprawling growth habit, they rarely have that much height unless they are clambering over other nearby plants.

Asiatic dayflower's seed capsules look like one side of a heart shape. Inside are one or two seeds that look a bit like miniature peas or beans.

Don't let the common name confuse you: Dayflowers are nothing like daylilies, which are in a completely different genus (*Hemerocallis*). What the two have in common, however, is that each individual flower only opens for one day. If you head to your best *Commelina* patch after sundown there won't be any of the lovely flowers for you to gather. (The other parts will still be good, though.)

Sustainably Harvest

Asiatic dayflower is a widespread weed. However, if you want to encourage a particularly lush patch, just be careful not to yank out the plants' shallow root systems. Instead, break off the top several inches of each stem.

When harvesting, try to break off a stem close to the ground by gently bending it. If it bends but doesn't easily break, it will have a tough texture. Keep working your way up the stem until you reach that snap-off point I described in the Techniques section. From the snap-off point to the tip, the stems will be just as good as the other parts. Below that point, only harvest the leaves because the lower stems may be unpleasantly tough.

The little "peas," the immature, green seeds within their half heart–shaped sheaths, are one of the most delicious parts of Asiatic dayflower, but so tiny that they're not worth attempting to harvest in quantity. Enjoy them as a trail nibble, or "shell" however many of them happen to be on the stems you harvest and enjoy with the rest of the plant.

I'm not a huge fan of the texture of the seed sheaths and usually leave these out of my recipes even though they are technically edible (there's a difference between edible and good).

Eat

The leaves, succulent stems, flowers, and immature seeds are good to eat at any time during dayflower's growing season. They are great raw, but also worth trying steamed and stir-fried. For the best mouthfeel, the stems should be chopped before they are eaten.

You can preserve Asiatic dayflower greens (stems and all) by blanching, then chopping and freezing them.

Creamed Dayflower

Serves 2

Asiatic dayflower contributes both greens and "peas" to this gentle, satisfying dish. It makes a good side with roast chicken and root vegetables. But don't drive yourself crazy trying to harvest too many of the tiny green seeds—just use whatever shows up with the stems and leaves you pick.

INGREDIENTS

½ pound dayflower leaves and stems (tender top few inches of the plants only)

Dayflower "peas" (as many as the bunch of tender stems you picked carried)

1 tablespoon butter

2 tablespoons water

2 tablespoons light cream

2 scrapes on a grater of fresh nutmeg

1 teaspoon fresh mint, minced

Salt

Dayflower blooms for garnish (Optional; keep in mind that if you harvest late in the day, all of the flowers already will have withered. But the dish is still good without the flower garnish.)

INSTRUCTIONS

1. Remove and set aside the flowers. Remove the "peas" from their half-heart sheaths and set aside. Discard the sheaths. Chop the stems and leaves.

2. Melt the butter in a pan. Add the water and the dayflower leaves and stems. Cook over medium heat, stirring constantly, until the leaves are completely wilted, the stems turning tender, and the water evaporated.

3. Add the "peas" and cook for 1 minute more.

4. Add the cream and the nutmeg and stir for just a few seconds until the sauce thickens. Remove the pan from the heat and stir in the mint and salt to taste. Serve garnished with the flowers.

Birch

BETULA LENTA (BLACK BIRCH) AND *BETULA ALLEGHANIENSIS* (YELLOW BIRCH)

Birch trees are easy to identify, even in winter, and provide foragers with four different delicious ingredients. The leaves and twigs make an excellent hot drink, the inner bark can be used as flour, and the sap can be boiled down into a fantastic syrup. Black and yellow birches are especially prized, as they have a wintergreen odor. Black birch's odor is the more noticeable of the two.

Find

Look for birch trees near water. They are among the first plants to grow after a fire or other ecosystem disruption.

Identify

At any time of year, the outer bark of birch trees stands out. Depending on the species of birch, it may be a silver-gray, white, buff yellow, or black color with striking dark, horizontal slashes. Up and down the trunk, narrow strips of the thin outer layer of bark loosen and peel horizontally into curling strips, giving the tree a shaggy appearance. The bark on older trees tends to darken and develop into scaly plates. The trees can grow 30 to 50 feet tall, although they are often smaller.

Birch leaves range from oval to triangular, usually a combination of both. They are alternately arranged on the branches and have pointy tips and double-toothed margins. Mature birch leaves are 2 to 3 inches long.

Birch catkins, as the flowers are called, look like short rodent tails. (I know that doesn't sound cute, but it's what they look like.) Male birch catkins are about 1¼ inches long and droopy. The female catkins start out short and upright, but eventually elongate and dangle downward.

Under the thin outer bark of birches is the soft inner bark. If you peel back a strip of it and expose it to air, it will first be cream-colored, then turn a rusty hue.

Birch seeds grow as nuts or nutlets in cones.

One of the most important parts of yellow birch and black birch identification requires using your nose: when scratched, their twigs emit a pleasant, faint, wintergreen scent.

All birch trees are deciduous, losing their leaves in autumn and greening up again in the spring.

Sustainably Harvest

There's no need to hack away at a birch tree to harvest its minty leaves, twigs, and inner bark. The branches break off easily, and after a storm or even just a windy day you will usually find plenty of newly fallen branches on the ground.

The leaves and smaller twigs are easy to break off of those fallen branches. To harvest the inner bark, use a pocketknife. Come in at an angle, as if you were going to peel the branch. The part you're after is the soft layer in between the papery outer bark and the hard wooden core. You'll feel when your knife has passed through the soft cambium layer (inner bark) and hit the wood. Carve the inner bark from the wood in strips. Dry it, then grind it into flour. Keep it refrigerated in an airtight container until you want to use it.

Tapping birch trees for their sweet sap requires care. The basic process entails drilling a hole into the trunk of the tree and inserting a spile, a tube, or even just a sharpened twig into the hole. The sap runs along or through that inserted item and drips into a bucket, bottle, or other container. The sap is quite watery and has to be reduced by boiling to make syrup.

Detailed instructions for tapping birch and other trees is on page 43 as well as on some websites listed in the Resources section, but one extremely important thing to remember is that the diameter of the tree's trunk determines the number of tap holes you can drill without killing the tree. Birch trunks are frequently less than 20 inches in diameter, so usually one tap per tree is the maximum. Do not tap trees with a diameter of less than 12 inches.

Timing is everything when it comes to tapping trees for their sap. For a few weeks in late winter, sap flows copiously and has excellent flavor. After that, the sap flow slows way down and the flavor develops a musty, "off" taste. The timing varies slightly from year to year, but you'll know the time is right for tapping when daytime temperatures climb above freezing but the nights are still below freezing. Note that birch tapping season is usually just after maple tapping time.

Eat

Birch's gently fragrant leaves, twigs, and inner bark all make a warming tea that is especially welcome in winter.

The flavor of birch syrup is quite different from that of maple, but very good. Another difference is that it takes about twice as much birch sap as maple sap to make a quart of syrup. Expect to need 80 to 100 quarts of birch sap to boil down to one quart of syrup. (Yes, you read that correctly!)

You could skip the boiling down process and simply use the watery, lightly sweet sap as is. You can make beer and wine with it, or just enjoy it as a refreshing drink.

The inner bark can be dried and ground into an aromatic flour. Birch flour is as much of a seasoning as it is a baking flour, so I almost always combine it with milder ingredients, as in this shortbread.

Birch Bark Shortbread

Makes 16 pieces

The inner bark of birch branches gives this shortbread a delicate, faintly wintergreen-like flavor. That flavor is intensified if you use birch syrup instead of honey as the sweetener.

INGREDIENTS

1 cup room temperature unsalted butter

⅓ cup honey or birch syrup

2½ cups all-purpose flour

¾ cup dried, powdered birch inner bark (made following the instructions above)

continued

INSTRUCTIONS

1. Preheat the oven to 300°F.

2. Cream the butter and honey or birch syrup together in a stand mixer or food processor.

3. Add the flour ½ cup at a time, incorporating each addition completely before adding more.

4. Add the birch bark ¼ cup at a time, again mixing each addition in fully before adding the next. If the dough becomes too stiff to work with, transfer it to a lightly floured work surface and knead the birch in by hand.

5. Pat the dough into a shortbread mold or a 9-inch pie pan. Deeply score it with a fork into 16 wedges.

6. Bake for 20 minutes. Crimp aluminum foil around the rim of the pie pan to prevent the edges of the shortbread from burning. Return to the oven for another 15 to 20 minutes.

7. Let cool for 10 minutes before removing from the pan and cutting along the previously scored lines into 16 triangular pieces.

HOW TO MAKE BIRCH SYRUP

There are two main methods for tapping a tree for syrup. One is similar to the way it's done commercially, while the other requires little more than a twig and a bottle.

For the first method, you drill a hole using a $^7/_{16}$- or $^5/_{16}$-inch bit. Drill at a slight upward angle and go in about 2 inches deep. Insert a metal spile (you can order them online—see Useful Resources appendix) into the hole and tap it in securely with a hammer or rock.

Once the spile is in place, you can attach plastic tubing to it and let the sap flow down into a bucket on the ground or a bottle tied to the tree. Or you can attach a hook to the spile and hang a lightweight bucket from the hook.

Be sure that all of your equipment is clean. It is a good idea to sterilize it with a solution of one part bleach to 20 parts water, and then rinse well with plain water.

For the second, more low-tech method, find a sturdy twig about ¼-inch thick and 2 to 3 inches long. Whittle one end down to a sharp, flat point. Place the sharpened end of the twig against the tree's trunk at an upward angle (the sharpened end will be pointing upward, the blunt end down toward you). Use the handle of the knife, a rock, or a hammer to tap the twig into the tree approximately an inch deep. Sap will start to flow out of the tree and along the twig. If it runs down the tree's bark instead, try carving a narrow groove along the length of the twig. Place a plastic water bottle or a lightweight pail directly under the twig, which is now acting as a spile. Attach the container to the tree with string or duct tape, tying around both the container and the tree. The twig spile will only transport a slow drip of sap compared to the previous method of tapping, but if you check your bottles and pails every day, you'll eventually have enough sap for a small batch of homemade syrup.

With either method, pass the collected sap through cheesecloth or another type of filter to remove any debris.

While you are stockpiling sap until you have enough to make syrup, keep it in the refrigerator or someplace equally cold. Sap will spoil if not kept cold and used within a week.

Once you've collected the sap, boil it in an open container so that the water content can evaporate quickly and the liquid can reduce to a syrup. Use a wide pan or the largest pot you've got, and don't pour in more than an inch or so of sap at a time. If this means you have to boil your sap down to syrup in several batches, so be it. In the end it will take less time and save you from watching your hard-won birch sap foam over the top of the pot. Lots of steam will come off the boiling sap, which is why this step is usually done outdoors.

When the steam coming from the boiling sap quiets down, transfer the sap to a smaller pot. It will still be very liquid at this point, but there won't be as much steam and you can finish the boiling off indoors if you like. It is still important not to fill the pot with much more than an inch of syrup-to-be because of the risk of foaming over.

When your syrup is hot it will be thinner than when it has cooled. To tell when you've boiled it long enough, keep a plate in the freezer while you're boiling the sap (or simply outside if it's cold enough). Pour about ¼ teaspoon of the syrup onto the frozen plate. Wait 30 seconds. Run a finger through it. When it has the consistency of syrup, it's done.

Blackberry
RUBUS ALLEGHENIENSIS,
R. FLAGELLARIS, R. CUNEIFOLIUS,
AND OTHERS

When I teach foraging I often start out by asking, "Who here has foraged for edible wild plants before?" Often only a couple of hands are raised. Then I ask, "Didn't you ever pick blackberries when you were a kid?" and suddenly almost everyone raises their arm high. I hope this entry will make it easier to find this familiar and beloved fruit.

Find
Blackberries ripen in late summer. They grow in full sun or partial shade, and are often found at the edges of woodlands, including the outlying borders of otherwise cultivated properties. They are also sometimes found in city parks.

Identify
Blackberry's oval, toothed leaflets (three to seven leaflets per compound leaf) have white or very pale green undersides (the tops of the leaflets are green). When the wind flutters those leaflets, those light undersides are noticeable even from a distance.

I should tell you to wear long sleeves because the arching, prickly canes could scratch your bare arms, but I confess I am guilty of rarely following that advice. It's hot out when blackberries are in season, so I'm usually in shorts and a tank top when I find them. But you get it that blackberries grow on prickly canes, right? Those canes frequently turn into a formidable hedge. If you're going after blackberries on purpose, you might want to dress for the occasion by wearing heavy blue jeans. A thick glove on your nonpicking hand can move canes out of your way. It also makes it easier to bring them closer to you when you need to reach for a good-looking cluster.

There are numerous species in the genus *Rubus* with the common name blackberry. These include *R. flagellaris* (dewberry), which grows low to the ground rather than in thickets of canes. Other than its prostrate growth habit and smaller (also edible) fruit, it resembles *R. allegheniensis* and other "brambleberries." Brambleberry is a catchall name for blackberries, raspberries, and other edible *Rubus* fruit.

Blackberry flowers are up to an inch in diameter

and have five white petals and numerous pollen-tipped stamens in the center.

Wild blackberry fruits look exactly like cultivated blackberries except that sometimes they are smaller. They look like lots of little bubbles pasted together (technically called an aggregate fruit). Together these "bubbles" add up to a ¾- to 1-inch-long fruit. When you pick a blackberry the fruit doesn't separate from its core the way raspberries do (in other words, raspberries have a hollow, thimble shape, but blackberries don't). Unlike similar-looking mulberries, blackberries don't grow on trees, and they will detach without any piece of stem.

Because blackberries do not all ripen simultaneously, you will find unripe white, green, and red berries on the plants at the same time as the softer, ripe, "black" (very dark purple) ones.

Sustainably Harvest

You are not harming the plant by harvesting the ripe fruit. In fact, blackberry plants can be invasive. But if you want a patch to thrive, don't harvest more than 20 percent of the leaves or roots. I can't imagine a situation in which this would actually be an issue.

The difference between a ripe blackberry and an almost ripe blackberry is galactic. One has a perfect balance of acidity, sweetness, and aroma; the other is just sour and astringent. Only pick the berries that offer themselves without the slightest need of a tug. They won't have a hint of red, but will be so deeply colored as to warrant their common name of blackberry.

Gather blackberry leaves anytime during the growing season; ditto the roots.

Eat and Use

I guess I don't have to tell you about blackberry jam or blackberry pie, and if you haven't already tried it, I'm sure you can imagine blackberry syrup on your pancakes. And then there's blackberry wine, blackberry turned into a sauce for pork or venison, dried blackberries, blackberry jelly, blackberry trifle . . . or just eat them right out of the patch on some late summer afternoon when the prickers don't mean as much as that next ripe berry.

Blackberry roots, fresh or dried and decocted into a strong extraction, are a folk remedy for diarrhea.

Blackberry leaves, fresh or dried, make a passable tea that is improved with the addition of an aromatic such as mint.

Blackberry Cobbler with Rose Geranium Syrup

Makes 6 to 8 servings

This recipe will still be delicious even if you don't have a rose geranium plant handy. But if you do, it will be heavenly.

INGREDIENTS

¼ cup rose geranium syrup or simple sugar syrup (instructions follow)

6 cups fresh or frozen blackberries (if using frozen, measure before defrosting)

3 tablespoons cornstarch, kuzu (kudzu) starch, or arrowroot powder

1 tablespoon lemon juice

½ teaspoon allspice

1½ cups all-purpose flour

1 cup sugar

¼ teaspoon salt

¾ cup butter (12 tablespoons or 1½ sticks), melted and then cooled for 5 minutes

INSTRUCTIONS

1. To make the rose geranium syrup, combine 2 cups sugar with 1 cup water and two rose geranium leaves in a small pot. Cook over medium heat, stirring constantly, until the sugar has dissolved. Cover so that the rose geranium essential oils that are flavoring your syrup don't evaporate, and let sit at room temperature for 1 hour. Pick out the rose geranium leaves. (Save the rose geranium syrup that will be left over from this recipe in the refrigerator. Try it drizzled over fresh berries and other summer fruit—divine!)

2. Preheat the oven to 375°F.

3. Combine the berries, cornstarch, lemon juice, allspice, and rose geranium syrup (or simple syrup)

If you don't have rose geranium leaves, you could substitute a few sprigs of lavender, or simply leave out the aromatic herb altogether and just make a simple syrup. It will still be good.

in a large bowl. Stir gently so that you combine the ingredients well without crushing the blackberries.

4. Transfer the fruit to a 9-by-9-inch baking dish. Make sure the fruit doesn't fill the baking dish more than ¾ full because you need to leave at least an inch of room for the topping

5. In a bowl, whisk together the flour, sugar, and salt.

6. Add the melted butter to the flour mixture (be sure to let the butter cool a bit before adding it to the other ingredients). Mix with your clean fingers

continued

until it forms a crumbly dough. It shouldn't form a solid ball, but the dough should hold together when you pinch a bit of it. If necessary, sprinkle 1 to 2 tablespoons of water into the dough, but err on the side of less rather than more water.

7. Grab a handful of dough and press it into a thick disk. The disk should be no more than ½-inch thick, but as thin as ¼ inch is fine. Lay the disk of dough (carefully—it will tend to fall apart) over one corner of the berry filling. Continue making palm-size disks of the topping dough and laying it over the other ingredients in the baking dish

until the blackberry filling is more or less covered. You can slightly overlap the disks of cobbler topping. If there's topping left over, crumble it over the top of the cobbler.

8. Put the cobbler on top of a baking sheet to catch the inevitable drippings. Bake for 45 to 55 minutes until the blackberry filling is bubbling and the cobbler topping is turning golden.

9. Remove from the oven and let cool for at least 15 minutes before serving. Blackberry cobbler is also excellent at room temperature or even the next day.

Blackberry Shrub

Shrubs are a way of preserving fruit juices via sugar and vinegar, and then using that to make refreshing beverages, both alcoholic and nonalcoholic. Shrubs were very popular centuries ago, and I'm happy to see them regaining some of that lost popularity because they are very versatile and surprisingly tasty. (I say "surprisingly" because most of us are unused to drinking vinegar.) No self-respecting forager who is also a mixologist would be without a shrub or two in his or her pantry.

INGREDIENTS

8 ounces fresh or frozen blackberries

8 ounces sugar

1 to 1½ cups apple cider or champagne vinegar

continued

INSTRUCTIONS

1. Mix together the blackberries and sugar. Cover and refrigerate for 24 hours. Stir well, then cover and refrigerate for another 24 hours. By this time the sugar will have drawn most of the liquid out of the blackberries.

2. Strain out the liquid. Measure the liquid and add an equal amount of apple cider or champagne vinegar (or better yet, blackberry vinegar if you happen to have some on hand).

3. Put the mixture into a clean glass jar, cover, and put into the refrigerator. Do not use for at least a week, longer if you can convince yourself to wait. Freshly made shrub is, in my opinion, too segregated in taste to be enjoyable. The vinegar and sugar jump out at your taste buds, muting the fruit flavor. But give it time (3 months is not too much), and the flavors harmonize, yielding an ingredient that is truly pleasurable to work with.

WHAT TO DO WITH BLACKBERRY SHRUB

Try simply mixing it with seltzer or other sparkling water. If you want to go for a cocktail, adding some blackberry shrub to dry sherry is good, with or without ice.

All of the "brambleberries," or fruits in the *Rubus* genus, are excellent in any blackberry recipe, including cobbler and shrub. These include raspberries, black raspberries, and wineberries.

Burdock

ARCTIUM SPECIES

Its deep taproots are both an excellent root vegetable and a medicinal powerhouse, and its flower stalks are a fantastic vegetable similar to the Italian vegetable cardoon.

Find

Burdock grows in full to partial sun in disturbed soils. Look for it in parks, unweeded areas of gardens and farms, alongside roads, and at the edges of fields.

Identify

Burdock is a biennial that starts its first year as a rosette of fuzzy leaves that are whitish on their undersides. These leaves can get to be quite huge, up to 3 feet long in *Arctium lappa*. Although the margins are untoothed, they are very wavy and look something like ruffles on a skirt.

The taproot, known as the vegetable *gobo* in Japan, grows straight down and looks like a long, light brown or beige carrot.

After overwintering, burdock sends up a flowering stalk from the center of its rosette of basal leaves in midspring. Learn to recognize that first, thick section elongating out of the center of the leaf rosette, because you want to harvest it before it reaches its next stage. That stalk will eventually get to be quite tall, anywhere from 2 to 9 feet. The flowers look very much like thistle flowers, purplish and shaped like round bottlebrushes.

Those flowers eventually become the burs, or seedheads, that give burdock its common name. They will stick to your clothes, your dog, and each other. They make handy camp fastenings if your shirt loses a button or a tent flap needs securing. George de Mestral, a Swiss electrical engineer, pulled burdock burs off his dog after a 1941 hunting trip in the Alps. Curious as to how the burs had such gripping power, he looked at them through a microscope. It took a decade for him to design, test, and develop a system for producing a machine-made version on two strips of nylon. In 1951, he applied for a patent on his hook and loop closures. The patent was granted in 1955, and Mestral eventually had a multimillion-dollar industry producing 60 million yards of Velcro a year.

Inside each of the burs are burdock's brown, crescent-shaped seeds.

Although by the time burdock flowers and goes to seed it is too late to harvest the root or stalk of that individual plant, it is worth learning to recognize it at that mature stage. Often last year's browned seed stalks will persist through winter, making it easy to spot from a distance where to look for this year's crop.

Sustainably Harvest

Burdock is considered an invasive weed, so sustainability is not an issue with this plant. If the root is the harvest you're after, look for the biggest rosettes of first-year plants. You'll know they are first-year plants because there won't be anything above ground except for that rosette of massive, ruffly leaves. If you're lucky, you'll find them in sandy soil or soon after a rainfall when they are easier to dig up. Bring a sturdy shovel or digging stick: These taproots grow straight down and deep. There is a direct correlation between size of the leaves in the rosette and the size of the root.

However, if you want burdock "cardoons" (the immature flower stalks), you have to let some of the plants reach the second-year flowering stage. For the thickest, best stalks, leave some of the bigger leaf rosettes alone. When the plants start to send up flower stalks in their second year, but well before they actually start to bloom, simply slice off the stalk near its base in the center of the leaf rosette. Usually the plants will try again and send up a second attempt at flowering, and even a third. I harvest the first and sometimes second round of stalks, but I leave the plants alone after that so they can seed future generations.

Eat

The roots are fantastic sliced and stir-fried with other vegetables or on their own. They have an earthy, musky flavor that can be lightened by peeling the roots before cooking them. For an even lighter flavor, soak the peeled roots in cool water for 20 minutes before draining and proceeding with your recipe.

The immature flower stalks need to be peeled because the outer layer is very bitter and stringy. As with their plant cousins, the artichoke and the cardoon, burdock stalks start to turn brown quickly when they are exposed to air. To prevent discoloration, have a bowl of acidulated water handy to put the peeled stalks into while you're prepping them (1 tablespoon of vinegar, 3 tablespoons of lemon juice, or ¼ teaspoon of citric acid per quart of water does the trick).

Gobo Stir-Fry

Serves 4 as a side dish, 2 as a main course

Pile this savory Japanese-style dish over rice or soba noodles. Add some tofu or poultry if you want to transform it into a substantial meal.

INGREDIENTS

½ pound burdock root
¼ pound carrots
1 tablespoon sesame seeds
1 tablespoon soy sauce
2 teaspoons honey
2 tablespoons mirin
1 tablespoon white wine
1 tablespoon vegetable oil

INSTRUCTIONS

1. Peel the burdock root and julienne it into matchstick-size strips. The peeling is optional. If you do peel the roots, you will have a milder dish. For a stronger, mushroom-like flavor, wash but don't peel.

2. Once all the burdock roots are sliced, drain off the acidulated water through a colander. Then put the burdock matchsticks in a bowl and let them soak in plain water for 30 minutes.

3. While the burdock is soaking, peel the carrot and julienne it into matchsticks as you did with the burdock root.

4. Toast the sesame seeds in a dry skillet over medium-low heat for a few minutes, shaking the pan often, until fragrant and just starting to color. Do not allow to burn.

5. Mix the soy sauce, honey, mirin, and wine.

6. Drain the burdock in a colander. Spread on a kitchen towel and pat dry.

continued

7. Put the vegetable oil in a frying pan or wok over high heat.

8. Add burdock and fry for 2 minutes, stirring.

9. Add carrots to the hot pan and fry for 2 more minutes, stirring constantly.

10. Stir the soy sauce mixture into the vegetables. Sprinkle with the sesame seeds.

11. Remove from heat and serve immediately.

Burdock, Bean, and Bacon Bake

Serves 4

You can swap coconut oil for the bacon to make a vegetarian version. It won't taste exactly the same, but it will still be very good. If you opt for the veggie version, consider adding ⅛ teaspoon of liquid smoke.

INGREDIENTS

3 ounces bacon, cut into small slivers

2 large leeks, tender parts only, washed and thinly sliced

¼ cup water

1 pound peeled immature burdock flower stalks

1 tablespoon garlic, minced

3 tablespoons butter

2 tablespoons flour

1 cup milk

1½ cups cooked white beans

1 tablespoon minced fresh sage leaves or 1 teaspoon crushed dried sage leaves

½ teaspoon salt

½ teaspoon freshly ground black pepper

⅛ teaspoon freshly grated nutmeg

¼ cup grated Parmesan or Romano cheese

½ cup panko breadcrumbs (okay to use other kinds of breadcrumbs, but the panko gives an especially nice texture)

continued

INSTRUCTIONS

1. Preheat the oven to 400°F. Lightly grease an 8-inch square baking dish.

2. Cook the bacon in a heavy skillet over low heat until most of the fat is rendered. Scoop out the bacon and set aside, but leave 1 tablespoon of the bacon fat in the pan. Add the leeks and water to the pan and cook, stirring often, until the leeks are softened and starting to caramelize. All or most of the water will have evaporated by then.

3. While the leeks are cooking, prep the burdock. Work with a large bowl of acidulated water nearby (1 tablespoon of vinegar, 3 tablespoons of lemon juice, or ¼ teaspoon of citric acid per quart of water). Peel the burdock stalks and cut them into ½-inch-thick pieces. Drop the burdock pieces into the acidulated water after peeling to prevent them from discoloring.

4. When the leeks are soft and starting to caramelize, add the garlic and cook, stirring, for 1 minute. Remove the pan from the heat.

5. Steam the burdock for 10 to 12 minutes until tender enough to easily pierce with a fork. Add the burdock to the other ingredients in the pan, along with the butter and flour. Cook over medium heat for 3 minutes, stirring constantly.

6. Raise the heat to medium-high and add the milk a little at a time, stirring constantly. Allow each addition of milk to thicken before adding more.

7. Gently stir in the reserved cooked bacon, beans, sage, salt, pepper, nutmeg, and half of the cheese. Spoon the mixture into a baking dish. Top with the breadcrumbs and the remaining cheese. Bake for 25 to 30 minutes until the top of the casserole is bubbling and turning golden.

Cattail

TYPHA SPECIES

This plant has something to offer foragers at any time of year. Its rhizomes, shoots, immature seedheads, and pollen are some of the most delicious wild edibles you can find.

Find

Where there are cattails there is a body of water, or at least some very soggy soil. Cattails grow in colonies with their roots in the sand or mud at the bottom of shallow freshwater. In rainy regions, they also sometimes grow in roadside ditches.

Identify

The seedheads that give this plant its common name look to me more like dark brown corndogs than the tail of any cat I've ever seen, but never mind. Sometimes called "punks," these seedheads stand upright on the tall (sometimes as tall as 9 feet) stalks well above the surface of the water the plants are in. The seedheads make it easy to identify cattails, even in the winter, when their rich brown color gives way to a scruffy white. Some foragers dry and grind the white fluff from the seedheads before they are fully mature to make a sort of flour. You also can dunk the punks in oil and use them for campfire torches, but they don't burn for very long.

The sword-shaped leaves are light green or blue-green. *Typha latifolia's* leaves are a bit wider than *T. angustifolia's* (don't worry—you can use both species interchangeably). Cattail leaves have been used as roof thatching and woven into baskets.

Foragers should be aware that yellow flag (*Iris pseudacorus*) has similar leaves and often grows alongside cattails. It is poisonous, as are all irises. Its leaves grow in a flat arrangement like the spokes of a fan, whereas cattail's leaves grow wrapped around each other at the base and facing each other.

Another plant with similar leaves that likes the same habitat is Sweet Flag, *Acorus calamus.* However, Sweet Flag has a spicy scent when crushed, whereas cattails don't really have a noticeable scent.

Cattail's flowers are cylindrical spikes. There is a male flower above the female flower. In the common cattail they touch; in the narrow-leaved cattail they are slightly separated. After it sheds its pollen

on the female flower, the male flower withers away, leaving the female to become the classic cattail "punk" or seedhead.

Cattails grow in colonies. If you poke around in the mud you'll find the maze of horizontal rhizomes from which the aerial parts of the plants emerge. Cattail rhizomes do not branch between plant stalks. The mature rhizomes are tan with stringy roots along their length, a spongy outer layer, and a more solid, whitish core.

Sustainably Harvest

Harvest the male flower heads while they are still green. Don't take them all, though. Leave some so that you can come back in late spring or early summer to collect the edible pollen.

Late fall through early spring is the best time to collect the rhizomes because that is when they are fat with the starch they are storing to fuel the next year of growth. Alas, this is also when the water temperatures in many places are freezing. I'm a lazy forager and limit my rhizome harvest to those cattails growing on land, which means I have to wait for winter thaws when the mud isn't frozen.

The starch from the mature rhizomes is good, but it's labor intensive to get and not as interesting to me as the other ingredients cattails provide. To harvest the mature rhizomes, use your hands to feel around in the mud. Yes, I know it's cold. I warned you.

Collect the lateral immature rhizomes after the plants have flowered in late summer and early fall. To harvest them, reach down into the mud and feel your way along one of the cattail rhizomes to its end. If the tip does *not* curve upward, you've found one ready to harvest. Go back from the tip to the point along it where you feel the first stringy roots (young laterals won't have any stringy roots). Just past that point, bend the growth end upward until it snaps off.

Harvest the shoots in spring. Use a knife to cut off the vertical green parts of the plants near the bottom. Another way of doing this, which doesn't require a knife, is to grab hold of most of the inner leaves and pull upward. The shoot will come loose. Peel down to the white, layered core. You'll feel like you're getting rid of most of the plant, but it's worth it. Your hands will get covered with a gel-like substance the plants exude. That gel is a good healing balm that is excellent for cuts and scratches.

Gather the pollen by bending the tops of the plants into a paper bag and shaking the plant inside the bag. It's best to do this on a nonwindy day or your harvest may blow away. In fact, the wind might *already* have blown your harvest away: If yesterday was windy, wait a couple of days before trying for a cattail pollen harvest. More pollen spikes may ripen.

Eat

The shoots are arguably my favorite cattail ingredient. They are wonderful steamed or stir-fried, with a mild taste and texture somewhat like hearts of palm.

The immature male flower heads look and taste something like the baby corn in a Chinese restaurant, except that the "cobs" are green instead of yellow. Before you get the wok out and toss your "baby corncobs" in with your stir-fry, steam or boil them first for the best texture. They have a slightly tough core. You can nibble the tender outer parts off separately, but the core doesn't bother me and I usually just eat the whole thing (something to be aware of, though, before you decide to serve cattail "corncobs" to guests).

You need to sift cattail pollen through a fine-meshed sieve or screen to remove any debris or insects. Use the pollen in combination with wheat or other grain flours in baked goods. Muffins, pancakes, and waffles made with cattail pollen replacing about 25 percent of the wheat flour are delicious and have a lovely golden color.

The immature lateral rhizome tips are also delicious, and fortunately not fibrous like other parts of the rhizomes. Simply wash them off and enjoy them raw, boiled, or steamed. They are also good in soups and stews, or glazed as in the recipe below.

Mature cattail rhizomes are so fibrous that they have been used to make cordage, which is not a tex-ture I usually find desirable in my food. But once you get rid of the fibers, the slightly sweet starch is useful as a flour- or cornstarch-like thickener.

To extract the starch from the rhizomes, first wash off any dirt. Peel the outer layer away from the cores. You will lose as much as half of the diameter of the rhizome, and the cores you wind up with should be white and smooth. Use your hands to break up the cores in a container of clean water. Twist, rub, and smash the cores to release the starch from the fibers. When you think you've gotten out as much starch as you can, remove the fibers. Let the starch settle to the bottom of the container. When it has, pour off the liquid, leaving the starchy goop in the bottom of the container. You can use this goop as is to thicken sauces and soups, or you can dehydrate it to make flour.

Store both cattail pollen and dried cattail starch flour in airtight containers. The immature lateral rhizomes can be blanched and frozen.

Buttered Cattail Shoots with Peas and Mint

Serves 4

This is a riff on the traditional English springtime dish of lettuce wilted in butter with peas and mint. The pleasingly mild flavor of the cattail shoots stands in for the lettuce here. Stick with just the whitest parts of the shoots for pure tenderness, or include some of the pale green bits if you want a sturdier dish.

INGREDIENTS

2 tablespoons unsalted butter

3 cups cattail shoots, chopped

½ cup water

1 cup fresh or frozen shelled peas
(if frozen, defrost them first)

2 tablespoons fresh mint, minced

Salt and freshly ground black pepper

INSTRUCTIONS

1. Melt the butter in a pot over medium heat. When the butter has melted, add the cattail shoots and water. Bring to a boil over high heat, then reduce the heat to low and cook, stirring often, until the cattail shoots are tender and most of the water has evaporated.

2. Add the peas and cook for 2 minutes more, stirring.

3. Remove from the heat and stir in the mint with salt and freshly ground black pepper to taste. Serve warm.

Glazed Cattail "Laterals"

Glazing is a perfect culinary technique to pair with cattail's delicious immature rhizome laterals. Glossy, tender, and lightly sweet, this dish is ready in just a few minutes. (You already put in enough time digging around in the mud for the laterals, right?)

It is crucial to use only the new, immature growth at the ends of the rhizomes and none of the extremely fibrous mature parts. See the Sustainably Harvest section above for how to make sure that you are getting the right part of the rhizomes for this recipe.

INGREDIENTS

1 pound cattail immature rhizome laterals, cleaned and cut into approximately ½-inch pieces

3 tablespoons butter

1½ teaspoons sugar

½ teaspoon salt

continued

INSTRUCTIONS

1. Put the cattail pieces into a pot small enough for them to be in one or at most two snug layers. You do not want them spread out in too large of a pot for this cooking technique. For the quantity in this recipe, an 8-inch pot is perfect. (But see below for some workarounds if you don't have the right size pot).

2. Add the butter, sugar, and salt to the pot along with enough water to just barely cover the cattail pieces (about 2 cups).

3. Bring to a boil over high heat. Continue to boil for approximately 10 minutes, shaking the pot occasionally, until the vegetables are tender and the liquid has reduced to a syrupy consistency. If the liquid is mostly evaporated before the cattail laterals become tender, add ½ cup water and cook a few minutes more. If the cattails are done but there is still a lot of liquid, remove them with a slotted spoon, boil off the liquid until it is syrupy, and then return the laterals to the pot.

4. Serve warm with additional salt to taste.

Chicken of the Woods/ Sulphur Shelf

LAETIPORUS SULPHUREUS

This is the gateway mushroom for many novice foragers. Bright yellow-orange and often growing halfway up a tree, it is easy to spot and hard to mistake for anything else. It is also one of the tastiest edible wild mushrooms when in good condition.

Find

Although there are several shelf mushrooms foragers call "chicken," including a mostly white one, *Laetiporus sulphureus* is the easiest to spot. It grows on oak and occasionally other hardwoods in eastern North America, primarily in summer and autumn, but occasionally in spring or winter as well. There are other similar looking and also edible *Laetiporus* mushrooms that grow in other parts of North America. The host trees vary from species to species. For instance, *L. gilbertsonii* is a West Coast species that grows on eucalyptus as well as oak.

Chicken of the woods mushrooms only grow on wood, and that is part of their identification. But I have seen chicken mushrooms that *looked* like they were growing on a lawn because there was a log buried just beneath the surface.

Identify

Laetiporus sulphureus fruiting bodies can grow nearly 2 feet across and are usually many individual caps arranged in a shelf-like or rosette formation.

These caps are up to an inch thick and have yellow pores on their undersides (there are no gills). The upper surface is bright orange or yellow in young specimens, fading with maturity and exposure to direct light. It has a suede-like texture.

There are no stems on chicken of the woods mushrooms. If you break one of the caps open, the flesh is white to pale yellow, moist in young chickens, becoming crumbly with age.

Sustainably Harvest

Although many mushrooms have essential, symbiotic relationships with trees, chicken of the woods is a parasitic mushroom that can eventually kill its host tree. You are not endangering the fungus by harvesting its fruiting bodies (the part that we eat). By the time you spot a chicken of the woods mushroom, its mycelium has already done irreparable damage to the tree. It can often be found on the stumps and logs of trees, as well.

Harvest by slicing off the mushrooms near the base. It's best to harvest them when they are young and tender, but there are uses even for the older, drier ones (see below).

Eat

Prime chicken of the woods mushrooms are young enough that their flesh is still moist and has a spongy but firm feel. They are absolutely delicious sautéed, fried, or simmered in stews. (Note that all wild mushrooms should be cooked.)

Older chicken of the woods mushrooms are often unnecessarily passed up by foragers. It's true that their chalky texture never becomes tender and moist, no matter how much liquid is in your recipe or how long you cook them. But once dehydrated and ground in an electric grinder, you've got a powder packed full of fantastic mushroom flavor that is excellent added to risotto and other dishes.

Chicken of the Woods Mushroom Pasta Sauce

Serves 4

This is a deliberately simple recipe because when I find a really choice, tender chicken mushroom I want to feature it, not overwhelm it with other flavors. You could, of course, use other mushrooms in this sauce, but the chicken mushroom is what turns simple into extraordinary here.

INGREDIENTS

4 tablespoons butter (or vegetable oil), divided

1½ pounds young chicken of the woods mushrooms, cleaned and finely chopped

1 shallot, peeled and finely chopped

¼ cup dry white wine or sherry

1½ cups milk (vegans can leave this out and double the amount of stock)

1½ cups mushroom or vegetable stock

3 tablespoons flour

Several sprigs fresh or ½ teaspoon dried thyme

Salt

Pepper

INSTRUCTIONS

1. Melt 1 tablespoon of the butter in a skillet over low heat. Add the mushrooms and shallot and cook, stirring often, until they have first released any liquid and then reabsorbed it, about 10 minutes. Add the wine and cook for another 5 to 10 minutes.

continued

2. Combine the milk and stock in a small pot and heat to a simmer.

3. In a separate medium-size pot, melt the remaining 3 tablespoons of the butter over low heat. Stir in the flour and cook, stirring often, for 4 minutes. Remove from heat and whisk in the hot milk/stock mixture a little at a time. (If you dump it all in at once, it will clump.)

4. Add the thyme, and return to the stove and simmer over medium-high heat for 5 to 10 minutes, stirring vigorously and often, until it starts to thicken. Add the mushrooms along with salt and pepper to taste.

5. Serve over any cooked pasta, but fettucine is especially good here. Go lightly with the grated cheese, if any.

Chicory

CICHORIUM INTYBUS

Chicory's edible leaves, roots, and flowers provide foragers with good ingredients across several seasons. You can purchase New Orleans–style coffee that is made with chicory roots . . . or you can easily make your own.

Find

Look for chicory in pastures, parks, and roadsides. Although it prefers full sunlight, I sometimes find it growing in partial sun.

Identify

Like dandelion, chicory regenerates from its perennial root system each year. Also like dandelion, it has deeply toothed leaves that initially grow in a rosette and that exude a milky sap when broken. Chicory's young leaves tend to be less pointed at the tip than dandelion's.

The similarities between dandelion and chicory end as soon as the plants send up flower stalks. Chicory's stalks branch and have small leaves on them, whereas dandelion's single (one per flower), hollow, flowering stalk never branches and never has leaves on it. Note that the leaves on chicory's flower stalks are smaller than those of the same plant's basal rosette, and that they often clasp the stalk.

Chicory's flowers are 1 to 1½ inches in diameter. The fringed ray flowers, which are the parts that look like petals, are usually an exquisite blue color, although they are occasionally pink. Chicory flowers only open for one day, and by the next day the withered blooms lose their color and can appear almost white.

The taproots are light brown on the outside and off-white inside.

Sustainably Harvest

Dig up chicory roots any time of the year for "coffee," but only in cold weather if you want to use them as a root vegetable. They are too bitter during the warm months of late spring through early fall.

The leaves and crowns (the crown is the area where the leaves attach to the root) should be harvested in early spring or late fall, when their bitterness is mild enough to be pleasant. Gather up the leaves of a chicory plant with one hand, while with the other you slip a knife just below the soil surface and slice off the entire rosette of leaves attached to a sliver of the root.

Pick the flowers just before you are going to serve them because they wilt rapidly.

Eat
Chicory's leaves and crowns are tastiest in early spring, when they are good both raw and cooked.

The lovely blue flowers are edible and make a good salad garnish if you serve them soon after harvesting. The roots make a tolerable cooked vegetable, but are at their best roasted and turned into a hot beverage. Add some coffee beans to roasted chicory root and you've got New Orleans' famous brew.

Big Easy Coffee

The classic coffee of New Orleans is a blend of roasted coffee beans with roasted chicory root. The mixture may have started out as a frugality issue—cutting expensive coffee beans with free chicory roots—but it persists because it is a delicious combination. And as a health bonus, chicory contains no caffeine, so Big Easy Coffee is "half caf."

INGREDIENTS

Chicory roots

INSTRUCTIONS

1. Use a sturdy vegetable brush to scrub a large heap of chicory roots clean under water. Finely chop them (the pieces should be less than ¼-inch thick or they will take a long time to dry and kill your grinder). Spread the chopped roots out on a baking sheet in a layer no more than ½-inch thick.

2. Roast in a 325°F oven. Stir occasionally so that the chicory roasts evenly. When the chicory roots are brittle-dry and as dark brown as you like your coffee beans (but not burnt), take the baking sheet out of the oven and let the chicory cool to room temperature.

3. If you grind your coffee just before brewing it, you might want to store your roasted chicory in its own jar. You can then simply add a spoonful or two to your grinder along with your coffee beans.

If you start your morning (or whenever) brew with preground coffee, I recommend grinding the roasted chicory in your coffee grinder and combining the two. You can go with equal parts coffee and chicory, or use slightly more of one or the other. Other plants you can use in this recipe: dandelion root.

Clover

TRIFOLIUM PRATENSE, T. REPENS, AND OTHER *TRIFOLIUM* SPECIES

Clover plants not only improve the soil in which they grow by fixing atmospheric nitrogen, their flowers provide delicious food and drink and potent herbal medicine benefits.

Find

Clover likes to grow in open, sunny spaces with disturbed soil. Look for it in your lawn, roadside ditches, meadows, alongside farms where it is often planted as a cover crop, and in urban parks.

Identify

Although other plants that also have leaflets in groups of three, including *Melilot* and *Oxalis*, are sometimes referred to as clovers, true clovers are in the *Trifolium* genus. *Trifolium* species are perennial plants that, in addition to leaflets in groups of three, have flowers that look like white, pink, or purplish pom-poms made up of many tiny individual florets. Most are in peak bloom in midspring through early summer, but can continue to put out occasional flowers all the way up until the first frost of fall.

Sustainably Harvest

Plants in the *Trifolium* genus are in the Fabaceae family, also known as the legume family. They are often planted by farmers as a cover crop because they are able to take atmospheric nitrogen and fix it in the soil, making that nitrogen biologically avail-

able to nourish other plants. Picking just the flowering tops of the plants allows the leaves and roots to keep doing their important work of photosynthesis and nitrogen fixing. But don't worry if a few leaves that hug the bases of the flowers end up in your collection basket: they are also edible.

Only harvest clover flowers that have no signs of browning, and only use them completely fresh or fully dried. They will taste better. Even more important, they will be less likely to contain harmful levels of a cyanide precursor, which clover can generate to protect itself after being damaged. ("Damaged" could include the stress brought on by cutting and wilting.) Fortunately, given adequate drying time, the hydrogen cyanide produced will evaporate from the clover. So again, remember to only eat completely fresh or thoroughly dried clover blossoms.

Eat (and Drink)

The flowers are the part of clover that are interesting as food. (Although the leaves are technically edible, they are not very tasty.) As mentioned above, the flowers should be used fresh or thoroughly dried. Mild and healthy in infusions, they are a versatile enough ingredient to also go into both grain salads and baked goods.

Medicinally, red clover (*Trifolium pratense*) is the most widely studied clover species. It is used for respiratory complaints and for chronic skin ailments such as eczema. Isoflavone compounds in red clover act as phytoestrogens and are used to relieve menopausal symptoms. There are several studies that indicate red clover may also be useful in preventing and treating breast cancer.

One of the best ways to use red clover both medicinally and as a pleasant beverage is to make an infusion of it. Red clover tastes mildly sweet to me and combines well with nettles, red raspberry leaf, and/or mint. To prepare, pour boiling hot water over the herbs, cover, and let steep for 30 minutes. Strain and serve hot or chilled. If you like your tea sweet, honey pairs better with red clover than sugar or agave.

You can strip the tender florets off their tough base and use them, fresh or dried, in grain recipes such as rice salads. Fresh red clover florets with barley and a little mint is an especially tasty combination.

Dried, the florets can be used to replace up to 25 percent of the wheat or other grain flour in recipes for baked goods. The red clover flowers add a lightly spongy texture, mild sweetness, and a dash of protein to whatever bread, muffin, etc. you are baking.

And here's one last fact about this fantastic plant: red clover is the state flower of Vermont.

Irish Clover Soda Bread

Makes 1 loaf

This soda bread recipe is ready to eat in less than an hour. The clover flowers add color, texture, a hint of sweet flavor, and all their medicinal and nutritional powers.

INGREDIENTS

1¼ cups whole-wheat pastry flour (pastry flour makes this bread more tender. If you can't get whole-wheat pastry flour, use a mix of half all-purpose and half whole-wheat flours.)

1 cup fresh or dried clover blossoms, measured after stripping them from the tough bases and separating them into individual florets

1 teaspoon baking powder

½ teaspoon baking soda

½ teaspoon salt

¼ cup butter, plus one more tablespoon reserved for brushing on the finished loaf

1 tablespoon honey

⅔ cup buttermilk or ⅔ cup milk whisked with 1 tablespoon vinegar

1 egg, beaten

INSTRUCTIONS

1. Preheat the oven to 375°F. Grease a baking sheet or line it with parchment paper or a Silpat mat.

2. Whisk the flour, fresh or dried red clover blossom florets, baking powder, baking soda, and salt together in a large bowl.

3. Warm the butter and honey in a small pot over low heat until the butter is melted and the honey is dissolved. Remove from heat, add the buttermilk first to cool the mixture, then the egg. Whisk to combine.

4. Pour the liquid ingredients into the dry ones. Stir to incorporate the flour. Don't stir too much, though—it's okay if there is still a little dry flour here and there; and for this dough, lumpy is good. You want the dough to still be somewhat soft and sticky, but coherent enough that you can shape it into a loaf. If the dough seems too goopy, add more flour a little at a time. Some cracks on top are okay and actually make the finished loaf more attractive in a rustic way.

5. Scrape the dough out onto your baking sheet. Shape it into a disk approximately 5 to 6 inches in diameter. Bake 25 to 35 minutes until golden.

While still hot, brush with the remaining table-spoon of butter. Let cool on a rack before slicing and serving.

Clover Flower Spoonbread

Serves 4

Spoonbread is a cross between cornbread and a soufflé. It is creamy enough to warrant eating with a spoon, hence the name. In this version, clover blossoms add a dash of color, texture, and a subtle but interesting layer of flavor. *Trifolium pratense*, *T. repens*, *T. hybridum*, and *T. clypeatum* are all equally good in this recipe.

INGREDIENTS

2 cups milk, divided

1 tablespoon butter

2 teaspoons honey (ideally clover blossom honey)

½ cup cornmeal or polenta

¼ cup fresh or dried clover blossoms, measured after stripping them from the tough bases and separating them into individual florets

½ teaspoon salt

1 egg, white and yolk separated

¼ teaspoon baking powder

INSTRUCTIONS

1. Preheat the oven to 350°F.

2. Heat 1½ cups of the milk plus the butter and honey in a medium-size pot, stirring occasionally until the butter and honey are completely melted.

3. Meanwhile, combine the cornmeal, red clover, and salt in a bowl. Add the remaining ½ cup of milk and stir to combine.

4. Whisk the cornmeal and clover mixture into the hot milk mixture a little at a time. Bring to a boil, then reduce the heat and simmer for 5 minutes, stirring constantly.

5. Lightly beat the egg yolk. In a separate bowl, beat the egg white with an electric beater until it forms stiff peaks.

6. Add a spoonful of the hot cornmeal and clover mixture to the egg yolk and stir it in. Add another spoonful and do the same, then add the tempered egg yolk to the pot of hot cornmeal-clover mixture.

7. Add the baking powder to the cornmeal-clover mixture and stir well. Gently fold in half of the

beaten egg white, then the remaining half. It's fine if the egg white isn't completely mixed in; in fact, you want to see pockets of fluffy egg white in the mix.

8. Spoon the mixture into lightly greased ramekins, filling them no more than ¾ full. Bake until puffed up and starting to turn lightly golden on top, approximately 20 minutes. Serve immediately.

Common Chickweed
STELLARIA MEDIA

Both raw and cooked, chickweed is a mild-flavored green that is quite cold hardy and in some regions can be harvested year-round.

Find

Chickweed loves the disturbed soils of gardens, farms, and parks and so is more likely to be found near human habitations than deep in the wilderness. It will grow in full sun or part shade, but chickweed in direct sunlight tends to be too stringy and ground hugging to be worth harvesting. Look for the more tender, lush patches that you will find in partial shade.

Identify

Common chickweed is a low-growing plant. It hugs the ground when it grows where it is exposed to full sun, but becomes a tangled mat of tender leaves and stems up to a foot tall and twice as wide when it grows in partial shade and moist soil.

The leaves grow in an opposite arrangement (they join the stem in pairs) with short or almost absent petioles (leafstalks). Each leaf is small, usually about half an inch long, but occasionally a full inch or sometimes as small as a quarter inch. The leaves are oval with pointy tips and smooth margins.

A unique characteristic of common chickweed is that if you hold a sprig of it up to the sunlight (or look at it through a magnifying lens) you will see a single line of hairs on the stem. Not uniformly

hairy stems, but just that one line of hairs, like a Mohawk hairstyle. The flower buds, on the other hand, are hairy all over. Note that there are other chickweed species, also edible but not quite as good (in my opinion), that are hairy over most of their aerial parts.

The ⅛-inch-diameter flowers look something like diminutive daisies, with their five white petals that are so deeply cleft that they look like 10 petals.

Under the petals of each flower are five green, hairy sepals that form a star shape.

Chickweed is primarily a cool weather plant, at least as far as good eating is concerned. In summer's heat its stems often become too stringy and to bother with. Even then, though, you can sometimes find tender, partially shaded patches of it that are worth harvesting. But mainly look for it in fall and spring, and sometimes even in winter if it's not buried under snow.

There is a poisonous plant called spotted spurge (*Euphorbia maculata*) that novice foragers sometimes confuse with chickweed. It is also a low-growing plant and likes similar habitats. The easy way to make sure you've got chickweed, not spotted spurge, is to break off a stem. If it is spurge it will ooze a white sap, but if it is chickweed it will not.

Sustainably Harvest

Chickweed is a European plant that has become an invasive weed in many places. Sustainability is not an issue with this excellent edible.

Look for a lush patch of chickweed growing several inches high rather than flattened on the ground. Gather a bunch of the stems in one hand, while with the other you either twist off or snip off with scissors the top 2 or 3 inches of the plants. All the aerial parts of chickweed are edible, including the flowers.

Giving chickweed a "haircut" encourages even more lush growth, and in this way you can harvest repeatedly from the same patch.

Eat

Unlike some other wild greens, chickweed is rarely bitter. It is one of my favorite salad greens. You also can use it in combination with other plants to make wild greens pesto. Or use it in any recipe to replace spinach as I've done in the hortopita recipe below.

There are some people, though, who *do* find chickweed bitter. Or maybe soapy would be more accurate. These are the same people who dislike cilantro (coriander leaves), which they also find soapy tasting. What is going on is that both cilantro and chickweed are high in saponins, which are, well . . . soapy. Some people's taste buds are more sensitive to saponins than others. If you don't like cilantro, you might not like chickweed, either. I'm glad my taste buds don't have this particular sensitivity, and that to me chickweed just tastes mild and good!

Hortopita

Makes 8 servings as a side dish, 4 as a main course

Many people are familiar with spanakopita, the Greek spinach and cheese pie with a flaky phyllo crust. This is that same pie, but made with chickweed and other wild greens instead of spinach.

Horta is the Greek word for wild greens and can be applied to any wild, edible leafy green vegetable. In this recipe, mild-flavored chickweed makes up the bulk of the horta, but I've added tangy dock and sorrel leaves for interest. If you choose to use only chickweed, add a little extra lemon zest to the filling.

INGREDIENTS

2 quarts tender chickweed (leaves, stems, and flowers)

1 pint any combination of the following: wood sorrel, sheep sorrel, or curly dock (leaves)

1 large onion, finely chopped

1 tablespoon olive oil

2 garlic cloves, peeled and minced

⅓ cup sour cream, cottage cheese, ricotta cheese, or Greek yogurt (do not use regular yogurt unless you strain it first or you'll end up with a runny pastry filling)

⅓ cup crumbled feta cheese

¼ cup grated Romano or Parmesan cheese

2 eggs, beaten

2 tablespoons minced fresh dill or 2 teaspoons dried dillweed

½ teaspoon grated lemon zest

½ teaspoon freshly ground black pepper

⅛ teaspoon freshly ground nutmeg

Salt to taste

¼ cup melted butter

Phyllo pastry dough

INSTRUCTIONS

1. Put 1 inch of water into a medium-size pot and bring it to a boil. Meanwhile, wash the chickweed and other greens. Add the greens to the water and boil, stirring, until fully wilted, not longer than 5 minutes. Note that the sorrel and dock will turn a dingy khaki color; this is normal and does not

eat. (You can use the same pot you cooked the greens in.) Add the garlic and cook, stirring, for 2 minutes more.

5. In a large bowl combine all of the ingredients except for the butter and phyllo pastry. Add salt to taste (you might not need any if your feta is especially salty).

6. Brush a 9-by-9-inch baking pan with some of the melted butter. Lay in a sheet of phyllo, brush that with more butter, and repeat the layers two or three more times. Note: keep the phyllo sheets wrapped in a moist, clean kitchen towel between adding each layer because they dry out *very* quickly.

7. Spread the chickweed mixture over the buttered phyllo layers.

8. Top with several more layers of phyllo, coating each layer with melted butter. Use the tip of a sharp knife to score the hortopita in a tic-tac-toe pattern. This allows steam to vent while it cooks and makes it easier to cut into individual portions later.

9. Bake until starting to turn golden, 30 to 35 minutes.

affect the flavor. The chickweed should retain its bright green color.

2. Drain the greens in a colander, then immediately rinse them under cold water to prevent residual heat from continuing to cook them. Drain again, then squeeze out as much liquid as you can (squeeze hard).

3. Finely chop the cooked greens. It will be easier now that they are in a tightly squeezed wad. If you had chopped the greens before cooking them, you'd have lots of little green bits stuck to the pot—much better in this order!

4. Preheat the oven to 350°F. Meanwhile, sauté the onion in the olive oil for 5 minutes over medium

You can serve hortopita hot, but in Greece it's more likely to be served at room temperature. (Translation: You can make this several hours ahead of serving.)

Other plants you can use in this recipe: Any leafy green that is good cooked will work in hortopita, but especially tasty are Asiatic dayflower, chicory, dandelion, garlic mustard, lady's thumb, lamb's quarters, mustard, quickweed, sow thistle, and wild lettuce.

Dandelion
TARAXACUM OFFICINALE AND OTHER *TARAXACUM SPECIES*

When I tell people that dandelion is my favorite plant, they often look incredulous . . . until I explain that this easy-to-identify plant provides at least four different vegetables, a hot beverage, wine, and three kinds of herbal medicine.

Find

You will find dandelion growing in open, meadow-like situations (as every lawn owner knows). It is also fond of city street tree pits, cracks in the sidewalk, neglected lots, and roadsides.

Identify

Dandelion leaves grow in a circular rosette arrangement, with all of the leaves connecting to the root at a central point. The leaves have jaggedly toothed or sharply lobed edges with the points facing either straight out or back toward the base of the leaf (never toward the tip). These points can look like fangs, which may be where the name dandelion comes from (the French version is *dents de lion*, meaning "teeth of the lion"). When growing in partial shade or in cultivation, the teeth are not as well defined.

There are other plants with similarly shaped leaves that grow in a basal rosette. Some of them, like sow thistle, even have similar-looking yellow flowers, and some have in common with dandelion a milky white sap that oozes out when you tear the plant. But here's how to verify that what you've got

is truly a dandelion: the flower stalks are *leafless* and emerge from the center of the *unbranched* leaf rosettes.

Dandelion's sunny flowers are up to 1½ inches in diameter. The flower buds first appear nestled down in the center of the leaf rosette, then shoot up on hollow, leafless stalks. Each stalk has only one flower head, and each flower head is made up of numerous small ray flowers that look like petals.

The flowers turn into the globular seedheads made up of dozens of wispy seeds. Although I loved blowing on these as a child, as an adult I confess they are the only part of the plant I don't have a use for.

The fleshy but slender, branching roots are brown on the outside and off-white within.

Sustainably Harvest

Dandelion needs no help from you to survive and propagate itself. It spreads by windblown seed (yes, you were propagating it when you blew on those seedheads as a child). It is also a tough perennial that will regenerate if you leave even a small chunk of its root system in the ground. In other words, don't worry about overharvesting this wonderful plant.

Collect the leaves in early spring and late fall, when temperatures are cool but not much below freezing. I like to slip a knife just below the soil surface and cut off a sliver of the root—that way the entire leaf rosette comes out in one piece.

Harvest the flowers by pinching them off near the base of the calyx (the green part holding the rest of the flower together). The main flush of dandelion flowers is in midspring, although there will be occasional blooms after that. But if you need a major dandelion flower haul for a project like dandelion wine, save yourself much frustration and do it during the peak spring bloom season.

You can use the roots at any time of year, but if it's the medicinal benefits you're after, the best time to collect them is when they are at their bitterest and fattest in early fall.

Eat

Enjoy early spring and late fall dandelion leaves and "crowns" (base of the leaf, top of the root area) raw or cooked. They are excellent barely wilted in stir-fries and can also be made into tempura.

Dandelion flowers can be made into wine, jelly, vinegar, or fritters, or chopped and added to omelets, muffins, even pancakes.

Dandelion roots make a passable cooked vegetable in soups and stews, but I think that's the least interesting use of this plant's many edible ingredients. Much tastier is to roast the roots and grind them for a caffeine-free, coffee-like hot beverage (use the method for Big Easy Coffee on page 69). The drink tastes good and has all the liver- and kidney-cleansing and digestive system–soothing medicinal properties for which dandelion is famous.

Wilted Dandelion Greens with Hot Bacon Dressing

Serves 2

INGREDIENTS

1 teaspoon extra-virgin olive oil

2 ounces bacon, cut into small pieces

1 teaspoon garlic, minced (wild or cultivated)

½ pound dandelion leaves, harvested before the plants flower

2 tablespoons apple cider vinegar

1 tablespoon honey

Salt and pepper to taste

INSTRUCTIONS

1. Heat the oil in a large skillet over low heat. Add the bacon and cook, stirring occasionally, until most of the fat has rendered out. Add the garlic and cook, stirring, for 30 seconds.

2. While the bacon is cooking, wash the dandelion greens. Spin them dry in a salad spinner or roll the leaves up in a clean dish towel to dry them. Chop the washed, dried leaves into approximately 2-inch pieces.

3. In a cup or small bowl, whisk the vinegar and honey together until the honey dissolves.

4. When the bacon and garlic are done, add the dandelion greens to the pan. Immediately turn off the heat and stir in the vinegar-honey mixture. Stir to coat the greens with the dressing and slightly wilt them. Add salt and ground pepper to taste. Serve while the dressing is still warm.

Yard Squid Tempura

Serves 4 as an appetizer

In very early spring, when dandelion greens are at their most delicious, my favorite way to harvest them is to gather the leaves of the small (at that time of year) rosette in one hand. With the other hand, I slip a knife just below the surface of the soil and slice off the top of the taproot. The result is a sliver of edible root holding together a whole rosette of young dandelion leaves. The result looks something like a vegetable version of calamari, hence the nickname "yard squid."

INGREDIENTS

16 "yard squids" (small, early spring dandelion leaf rosettes with a sliver of their root system holding them together), scrubbed and rinsed clean

3 tablespoons soy sauce

3 tablespoons mirin

1 tablespoon rice vinegar

1 teaspoon dark (toasted) sesame oil

½ teaspoon grated fresh ginger

2 to 3 cups vegetable oil for frying

6 tablespoons all-purpose flour

1 tablespoon cornstarch

½ teaspoon salt

1 cup very cold sparkling water (club soda, seltzer, etc.)

INSTRUCTIONS

1. After washing, pat the "yard squids" dry with a dish towel.

2. For the dipping sauce, whisk together the soy sauce, mirin, vinegar, sesame oil, and ginger.

3. Pour enough of the vegetable oil into a pot or wok to fill it to a depth of at least 1 inch. Heat the oil

continued

until a drop of water sprinkled on it immediately evaporates, but not so hot that it smokes.

4. Whisk together the flour, cornstarch, salt, and cold sparkling water. Immediately after combining the batter ingredients, dip each dandelion crown in, shake off excess batter, and drop into the oil. Do not add too many "yard squids" at a time—you don't want to cool the oil down too much. Fry for about 2 minutes until golden, then remove with a slotted spoon and drain on a paper towel–lined plate. Repeat with the remaining dandelions. Serve with the dipping sauce while still hot. (You can keep the "yard squids" that are finished warm in a low oven while you fry the others.)

Other plants that work well in this recipe: chicory, sow thistle, and wild lettuce.

Dandelion Beer

Makes about 12 bottles of beer

INGREDIENTS

½ pound dandelion leaves, washed and chopped

½ ounce roasted dandelion root

½ ounce fresh ginger, grated (no need to peel)

5 quarts water

1 tablespoon beer yeast (you can get away with using bread yeast if necessary)

1 pound raw or Demerara sugar

1 ounce cream of tartar

¼ cup warm water

INSTRUCTIONS

1. Put the dandelion and ginger into a large pot with the water. Bring to a boil over high heat. Reduce heat and let it cook at a lively simmer for 10 minutes.

2. While the dandelion is brewing, mix the yeast with ¼ cup warm water in a small cup or bowl and set aside.

3. Put the sugar and cream of tartar into a clean fermentation vessel (you can use a food-safe bucket or even a large stainless steel stockpot but don't use aluminum).

continued

4. Line a colander with a double layer of cheesecloth or butter muslin. Put the colander over the sugar and cream of tartar in the fermentation vessel and pour the dandelion-ginger brew through it. Remove the colander. Stir to dissolve the sugar.

5. Let the brew cool to room temperature before stirring in the yeast. Cover the vessel with a dish towel and set in a warm but not hot place (warm-ish room temperature is good). Leave for 3 days, stirring vigorously at least once a day.

6. Siphon the beer into sterilized beer bottles and cap. Be careful to leave the lees (the yeasty guck at the bottom of the fermentation vessel) behind while you're siphoning.

7. Store the bottles on their sides in a cool place (your fridge or a cool garage will work) for at least 7 days before drinking. It will be even better if you wait 3 weeks.

Note: If you're new to home brewing and aren't sure how to go about things like siphoning, sterilizing, and capping bottles, please check the Resources section.

Dock

RUMEX CRISPUS, R. OBTUSIFOLIUS, AND R. PULCHER

From its large leaves in early spring to its tangy flower stalks and rust-colored seedheads, dock provides three substantial and easy to harvest ingredients.

Find

Dock plants—also known as curly dock, bitter dock, broadleaf dock, fiddle dock, and yellow dock—like to grow in full to partial sunlight and disturbed soil. Look for them in parks, farms, unweeded gardens, lots, and roadsides. They like water and often grow near rivers, lakes, and ponds. Like many other species in the *Rumex* genus, they are salt tolerant and so are also often found near saltwater shores.

Identify

Docks are perennials with large leaves that start out growing in a rosette arrangement. Those leaves are hairless and often have some red coloration on the midribs. Clasping the base of the leaves is a sheath called an ocrea, which is characteristic of docks. The ocrea starts out soft and a little bit slimy, then becomes dry and turns brown as the growing season progresses.

One of dock's common names is yellow dock, and that is because the roots are mustard yellow when you cut them open. The roots are the part of the plant that is used in herbal medicine, and they are considered one of the best remedies for problems with the liver and the digestive system.

The grooved flowering stalks emerge from the center of the basal leaf rosette. They grow to be several feet tall and have swollen nodes where the alternate leaves join the stalk. The leaves on the flower stalks are much smaller than the basal rosette leaves.

The clusters of brownish-green flowers become rusty-brown three-parted husks. Inside each husk is a seed. Each seed is a three-sided achene that often has a round tubercle on one or all three sides. You can spot the distinctive seedheads from a distance even in winter.

Curly dock (*Rumex crispus*) has narrow leaves that, although they are not toothed, are so wavy at the edges that they look ruffled. In botanical geek speak, those are called "crisped" margins, hence the species name *crispus*.

Broadleaf or bitter dock (*Rumex obtusifolius*) has less wavy, wider leaves that are shaped something like an oversized gardener's trowel.

Fiddle dock (*Rumex pulcher*) leaves are somewhere between curly dock and broadleaf dock with occasionally "crisped" margins. They sometimes veer toward an arrow shape with two lobes at the base, like a giant version of sheep sorrel, which is in the same genus. The flower clusters are much looser than those of *R. crispus* or *R. obtusifolius*.

Sustainably Harvest

Docks are considered invasive in many regions and sustainability is not an issue with these plants. Besides, if you are only harvesting the aerial parts (leaves, shoots, seeds), they will regenerate from the perennial roots.

Gather the leaves in early spring, when they are still in the rosette-only stage (no flower stalks, only leaves emerging from a central point where they connect to the root). Look for the newly unfurled leaves that have faint creases lengthwise, showing where they just unfolded. Those newly opened leaves will be the most tender and delicious.

Dock leaves become extremely bitter by summer, so it's best to harvest them during spring's cooler temperatures. Fortunately, just as the leaves start to become too bitter to be palatable, the next dock vegetable is ready to harvest. Look for the young flower stalks just as they begin to lengthen and well before they actually flower. Cut them off near the base.

Dock's rust-brown seed clusters are easy to spot from a distance in summer and fall and often right through winter. Cut off the whole seed clusters and let them dry out in paper bags or on screen-lined trays for a few days before stripping them off the stalks. Personally, I think it is too much trouble to separate dock's tiny seeds from their tenacious chaff, so I simply use them together.

Eat

During cool weather in the growing season, dock leaves are excellent raw or cooked. They have a slightly sour, slightly bitter taste and are fairly mild so long as the weather is cool enough. In warm weather they become too bitter for most people to enjoy.

The young flower stalks, harvested when they have not yet flowered, are an excellent vegetable if you're okay with mucilaginous textures. They have a delicious tangy flavor and are good raw or in soups, where that slight sliminess becomes a positive thing because it thickens the soup.

The seeds can be used in baked goods, but it's not worth bothering to winnow them. Just use them whole and call it extra fiber.

Dock of the Bay Fish Stew

Serves 6 to 8

This dish is based on cioppino, the seafood stew made famous by Italian immigrant fishermen in San Francisco in the late 1800s. You can vary the fish to use whatever is in season, as those fishermen no doubt did out of necessity.

INGREDIENTS

2 tablespoons butter

½ cup extra-virgin olive oil

2 medium onions, chopped

2 large celery ribs, chopped

1 large red, orange, or yellow bell pepper, stemmed, seeded, and chopped

3 garlic cloves, minced

2 pints tomatoes, chopped (canned or fresh)

2 pints fish stock or clam juice

2 bay leaves

½ teaspoon dried thyme leaves

½ teaspoon dried oregano leaves

1½ cups dry red or white wine

1½ pounds bay scallops

2½ pounds fish fillets (halibut, cod, or salmon), cut into bite-size chunks

½ pound dock leaves, chopped

½ cup fresh parsley, minced

¼ cup fresh basil, minced

Salt and freshly ground pepper to taste

continued

INSTRUCTIONS

1. Melt the butter into the olive oil in a large pot over medium-low heat. Add the onions, celery, and bell pepper and cook slowly, stirring often, until they soften. Add the garlic and cook for 1 minute more.

2. Add the tomatoes, fish stock or clam juice, bay leaves, thyme, oregano, and wine. Raise the heat to high and bring to a boil. Reduce the heat to low, partially cover, and simmer for 45 minutes. If the sauce becomes too thick, add a little water or additional wine.

3. Add the scallops, fish, and dock leaves. Cover and simmer for 10 minutes.

4. Remove from heat. Stir in the parsley and basil. Add salt and pepper to taste.

Massaged Salad with Radish and Pistachio

Serves 4

Rubbing sturdy leafy greens with an acid like vinegar and/or some salt softens them and yields a texture that is different from either raw salad or cooked greens. It is a technique most often used with kale, but it works just as nicely with young dock leaves.

INGREDIENTS

½ pound dock leaves, midribs removed and roughly chopped

¼ pound garlic mustard leaves or mustard greens

¼ teaspoon sea salt

4 tablespoons rice or cider vinegar, divided

1 tablespoon soy sauce or tamari

2 tablespoons tahini

⅓ cup radishes, sliced

½ cup scallions, chopped

2 tablespoons pistachio nuts, crushed

INSTRUCTIONS

1. Chop the dock and mustard greens. Put them into a large mixing bowl along with the chopped dock midribs. Sprinkle the greens with the salt and 2 tablespoons of the vinegar. Using your clean hands, vigorously massage the greens until they wilt.

2. Whisk together the remaining 2 tablespoons of vinegar, plus the soy sauce and tahini. Depending on the tahini you use, you might need to add a little water. The dressing should be the consistency of light cream.

3. In a small bowl, combine the dressing with the radishes and scallions and let them marinate for 10 minutes.

4. Toss the greens with the radishes and scallions. Sprinkle the pistachio nuts over the top and serve.

Dulse

PALMARIA PALMATA

Dulse is one of the easiest seaweeds to identify, and even people who are not fans of other sea vegetables often like the taste of this one.

Find

You can find dulse on most rocky shorelines on several continents, thought it isn't always attached to rocks; sometimes it attaches to other seaweeds. Although you can harvest it year-round, the taste is better during the colder months. But since you're more likely to find yourself wading at a beach shoreline in summer, go ahead and look for it then.

Identify

This plant's scientific name, *Palmaria palmata*, hints twice at one of the main identification characteristics of this algae. Dulse's fronds are one piece at the base, like the palm of your hand, but then split into finger-like extensions. That's where the hand analogy ends, though. Dulse fronds are flat, coming off a disc-shaped base called a holdfast, and frequently have "extra" leaflets coming out of the otherwise smooth margins, especially near the base.

Besides the hand shape and the leaflets emerging from the margins, the other characteristic that makes dulse easy to identify is its maroon to reddish color. Most other seaweeds are green, yellow, or brown.

Sustainably Harvest

Gather dulse in a location where it is frequently bathed by strong currents. It's easiest to collect when exposed by low tide. It is extremely important that the water at the harvesting location is clean. You can look up approved shellfish harvesting and fishing locations to verify that the waters have been tested and are unpolluted enough to harvest from.

Collect only the upper two-thirds by snipping them off with scissors. Leave the holdfast and the bottom parts of the fronds attached to the rocks so the dulse can grow back from them.

Eat

Fresh dulse isn't that great in my opinion; it's chewy and nondescript. But I know foragers who disagree with me on that, so go ahead and give it a try. Commercially sold dulse is usually put through a softening process that involves aging and enzymatic action similar to what is done with some cheeses.

What almost everyone agrees on is that dulse is delicious once you have dried it. That means drying it first, even if you are going to rehydrate it later before using it in a recipe. You can dry dulse by simply spreading it out in a single layer on a screen and putting the screen out in the sun or in a dehydrator at 110°F.

I enjoy dried dulse as is for a chewy snack, but an even better option is to make dulse chips by crisping dried dulse in a skillet with a little vegetable oil. You can also skip the oil and crisp the dulse in a dry skillet or low oven.

Cut dried dulse into thin ribbons or bite-size pieces and add it to salads, chowders, stir-fries, soups, and sandwiches. You can also grind it into a seasoning that will add a nutritional boost as well as flavor to anything you sprinkle it on.

Along with adding a pleasantly briny flavor, dulse will give you a good dose of iodine (essential for your thyroid) and B vitamins, especially B6. Dulse also has a high protein content.

Stored in airtight containers, dried dulse keeps indefinitely, although its vitamin B content will decrease over time.

Hebridean Dulse Broth

Serves 4

A rich-tasting Scottish recipe, Hebridean broth just needs to be served with a good, crusty bread for a satisfying meal.

INGREDIENTS

1 ounce dried dulse

½ pound white or yellow potatoes, peeled

1 tablespoon butter

1 teaspoon lemon juice, divided

3 cups milk

Salt and pepper

INSTRUCTIONS

1. Put the dried dulse in a bowl and pour water over it. Let it soak for 10 minutes.

2. Drain the dulse in a colander, then put it into a saucepan with enough water to cover and boil for 10 minutes. Add more water if any dulse starts to stick to the saucepan. Drain again.

3. While the dulse is cooking, chop the peeled potatoes into big chunks. Put them in a pot with enough water to cover. Bring to a boil over high heat. Cook until the potatoes are fall-apart tender when pierced with a fork, about 15 to 20 minutes. Drain.

4. Put the potatoes into a pot with the dulse, butter, and ½ teaspoon of the lemon juice. Mash with a potato masher or the bottom of a sturdy, heatproof bottle. Put the pot over medium heat and stir in the milk. Raise the heat to high and bring barely to a boil. Reduce the heat and simmer for 20 minutes, stirring frequently. Add salt and pepper to taste as well as the rest of the lemon juice if you think it needs it.

Garlic Mustard
ALLIARIA PETIOLATA

Garlic mustard is a gardener's bane because it is an extremely invasive plant. Fortunately, it is also an extremely tasty one with edible leaves, flowers, seeds, and roots. Eat the invasives! It may or may not bring them into balance with the rest of your local ecosystem, but at least you'll get the bonus of good, free food from your weeding efforts.

Find

Many common garden weeds prefer the combination of full sun and disturbed soil, but garlic mustard prefers partial sun or even partial shade. It frequently grows under deciduous trees, where it can get its early spring growth done before the trees leaf out. Garlic mustard is also common in the edges between gardens and woods and along property tree lines.

Identify

Garlic mustard is a biennial, which doesn't mean it takes 2 years to grow. It means it starts its growth during the summer or fall as a rosette of heart- or kidney-shaped leaves. Those first-year plants then overwinter, flower, and go to seed the following year.

Garlic mustard often shares its partially shady habitats under deciduous trees with wild violets. This can confuse beginners because both violet leaves and garlic mustard's first-year leaves are somewhat heart shaped. But instead of violet's pointy-toothed edges, garlic mustard's rosette leaves have softly scalloped margins. Instead of violet's

ridge-like veins, garlic mustard's veins form a net-like pattern. The colors are also slightly different, with garlic mustard leaves a yellow-green and violet leaves more of a blue-green. The flowers are distinctly different. But use your nose to clinch your ID by smelling a crushed leaf: violet leaves don't have a smell, whereas garlic mustard's smell strongly of, well, garlic and mustard.

After overwintering, garlic mustard greets the spring by shooting up flower stalks that can get to be 2½ feet tall. The flowers start out looking like miniature broccoli heads, then open into small, four-petaled white flowers. The leaves on the flower stalks, which grow in an alternate arrangement, are smaller and have a more pointed, triangular shape than the basal rosette leaves.

The flowers become slender, dry capsules 1 to 2½ inches long. Each capsule is filled with black or very dark brown seeds.

Sustainably Harvest

Garlic mustard is one of the most invasive plants around, and you don't have to worry about overharvesting it. In fact, be careful not to scatter the seeds around while you're harvesting them—this plant doesn't need any encouragement.

The absolute peak harvest time for garlic mustard, in my opinion, is when the plants are just sending up flower stalks in the spring. Maybe there are a few unopened flower bud clusters already showing on the plants, but the stems are still tender enough to snap off easily rather than just bending and bending, which is a sign that they will be too tough to eat. Harvest all of the aerial parts at this stage: stems, leaves, and flowers.

Once the skinny seedpods start to show, you can still harvest the leaves, but the stems may become too tough to be good eats. The green seedpods, however, are a very tasty treat in their own right. Break off the whole tips with several tender seedpods on them, then strip the seedpods off the stems.

Once the pods have turned tan and the dark seeds within are easy to shake out, it's time to collect the seed harvest. The best way I know to do this is to put a cloth or paper bag over the top(s) of the plant, bend the plant until the bag is upright, and then break it off. Dry the seeds for a few days, then crush lightly (still in the bag) with a rolling pin or wine bottle to loosen the seeds. Dump the seeds and chaff out into a big bowl and toss in front of a fan or a good wind to winnow. The seeds will drop back into the bowl and the chaff will blow away.

Garlic mustard roots should be harvested from first-year plants that have a leaf rosette only, no flower stalk.

Last but not least, those rosette leaves on the first-year plants can be slightly bitter, but they are available most of the year and still worth collecting to make chips or add to sandwiches instead of lettuce. They're okay mixed with other wild greens in a salad.

Eat

At the peak stage, when the plants are just starting to flower but the stems are still tender enough to snap off easily, eat the whole aboveground plant raw or cooked. Garlic mustard at this stage is fabulous in any recipe that calls for broccoli rabe.

In mid- to late spring, the still-green young seedpods make a pleasantly spicy trail nibble. And a great way to enjoy the basal rosette leaves is as chips following the recipe for plantain chips on page 189.

Garlic mustard roots have a good horseradish flavor, but unlike horseradish they never develop a stout taproot and are small and stringy. But if you mince them finely and preserve them in vinegar they make a good, spicy condiment.

Garlic Mustard Pesto on Crisp-Creamy Polenta

Serves 4

Wild food aficionados may roll their eyes when they see that I'm including this recipe because pesto is used as the go-to recipe for this plant so often that it's become cliché. But there's a reason for that: it's really, really good.

You can toss garlic mustard pesto with pasta, of course, but a spoonful added to soup just before serving is also wonderful, as is a smear of it on focaccia or toast. My favorite way to enjoy garlic mustard pesto is on pan-fried polenta that is crispy on the outside and creamy within.

INGREDIENTS

2 cups fresh garlic mustard leaves and tender stems

3 tablespoons walnuts or pine nuts, chopped

1 teaspoon garlic, minced (wild or cultivated)

¼ cup Parmesan or Romano cheese, grated

½ cup plus 2 tablespoons extra-virgin olive oil, divided

2 tablespoons butter

8 slices (½-inch-thick) cooked polenta

continued

INSTRUCTIONS

1. Put the garlic mustard leaves, nuts, and garlic into the blender or food processor. Pulse until the leaves are chopped.

2. Add the cheese. With the motor running, add ½ cup of oil a little at a time until the mixture is well blended but not completely smooth. (You want a bit of texture from the nuts and greens to remain.)

3. Heat the butter and 2 tablespoons oil in a large nonstick pan over medium-high heat. Add the polenta slices. (You can use the precooked polenta that comes out of a tube, or if you cooked some from scratch, spread it out ½-inch thick on a baking sheet and refrigerate until sliceable.)

4. Don't try to move the polenta slices until they've browned on the bottom side. You'll know that's happened when they dislodge easily. Use a spatula to flip them over and brown the other side.

5. Plate two slices per person, with the garlic mustard pesto spread on top. Serve hot or at room temperature.

Tip: If you want to keep this pesto in the refrigerator for up to a week or in the freezer for up to 6 months, blanch the garlic mustard greens in boiling water for 20 seconds, then immediately run them under cold water or dip them in an ice bath. Squeeze out as much water as you can, then proceed with the recipe. This blanching step prevents the pesto from losing its bright green color and turning brown in cold storage.

Simple Supper Garlic Mustard Pasta

Serves 4

This is a simple but satisfying one-pot meal that comes together in about 20 minutes total. You can embellish the recipe with additional ingredients such as chorizo sausage or pine nuts, but it's really not necessary. Sometimes simple is best.

INGREDIENTS

1 pound penne pasta

1 pound garlic mustard leaves and shoots, washed and coarsely chopped (ideally you're using garlic mustard at the stage where the stems are still tender and the flowers are either still in bud or just starting to open)

4 garlic cloves, peeled

1 to 2 medium-hot red chili peppers (pepperoncini), stems and seeds removed

¼ cup plus 1 tablespoon extra-virgin olive oil, divided (use your best because this is one of the main flavors of the sauce)

Salt to taste

½ cup Parmesan or Romano cheese, freshly grated (again, use the best you've got)

Freshly ground black pepper

INSTRUCTIONS

1. Bring a large pot of water to a boil. Add the penne and set a timer for 7 minutes.

2. While the pasta is cooking, prep the other ingredients: wash and chop the garlic mustard, mince the garlic or put it through a garlic press, chop the chili peppers.

3. After 7 minutes, add the garlic mustard to the pasta in the pot and cook until the pasta is al dente, usually about 5 minutes more.

4. Scoop out a ladleful of the pasta cooking water and set it aside. Drain the pasta and garlic mustard in a colander. Return the pot to the stove over low heat.

5. Add 1 tablespoon of the olive oil to the pot along with the garlic and chili pepper. Cook, stirring constantly, for 1 minute. Return the reserved pasta cooking water and the drained pasta and garlic mustard greens back to the pot. Raise the heat to medium and cook, stirring, for a minute or two until the liquid is mostly evaporated or absorbed. Remove from the heat, then stir in the remaining

continued

olive oil and salt (go scant on the salt because the grated cheese you'll be adding is salty).

6. Serve hot with freshly grated cheese and freshly ground pepper.

Other wild edibles you can use in this recipe include any leafy green as well as the leaves of any wild garlic species.

Hawthorn
CRATAEGUS SPECIES
Both a food and a medicine, hawthorn fruit makes beautiful preserves and liqueurs. But watch out for those thorns!

Find
Look for hawthorn on open hillsides, pastures, and near streams. It likes to grow in full or partial sunlight. Landscapers have planted hawthorn trees in many city and suburban parks.

Identify
Hawthorn trees are fairly small, growing between 10 and 30 feet tall. Their leaves are always alternate with toothed margins, but the shape can vary from species to species. Some are lobed, others almost oval.

The flat clusters of white to pale pink flowers bloom in mid- to late spring. They look a bit like apple or cherry blossoms. Each flower has five petals.

The fruits, which are the part you're going to harvest, look something like crabapples hanging in sparse clusters. They are usually red, but sometimes closer to burgundy. Unlike crabapples and other apples, which always have five seeds arranged in a pentacle pattern, the number of seeds in hawthorn fruit can vary from one to five.

The other thing that will help you distinguish a hawthorn tree or shrub from an apple tree at a glance is the thorns. These can be 2 inches long and sharp.

Sustainably Harvest
Picking the fruits does not harm the tree. Gather hawthorn fruits when they are fully ripe and starting to fall from the trees. You can collect already fallen fruit, but be careful to inspect your haul closely and discard any fruits that are bug infested or badly bruised.

Collect the leaf buds in early spring.

Eat
Hawthorn fruits are not especially interesting raw except as a trail nibble because of their often mealy texture. But they are excellent in any recipe that

highlights their flavor and the beautiful blush color the skins provide. Hawthorn jelly is lovely, as is hawthorn syrup. Hawthorn fruit is rich in pectin, the substance that causes jellies to gel, and you can combine it with low-pectin fruit such as strawberries for that purpose.

Tincturing hawthorn fruit in alcohol captures not only its taste and rosy hue, but its medicinal properties as well. It has a reputation as one of the best herbal medicines for the heart, normalizing blood pressure, and it is also useful for irregular heartbeats. Hawthorn is rich in powerful antioxidants, including proanthocyanidins and flavonoids. Numerous studies have shown that hawthorn not only makes arteries more pliable, but actually helps to repair damaged vessel walls. It is also useful for an upset tummy or diarrhea. In any case, a sip of hawthorn cordial is easy to take (for medicinal purposes only, of course).

Cooked, you can turn hawthorn into fruit sauce, fruit butter, or chutney just as you can apples. A food mill is handy for removing the seeds for any of those products, as it is for the hawthorn ketchup recipe, but with just a bit more work you can get away with using a sieve.

Hawthorn leaf buds can be added to salads. Both the leaves and flowers can be made into tea or syrup. For these purposes, I think it is best to dry the flowers first. Fresh, they can have a fishy smell due to the presence of trimethylamine, a substance that attracts pollinators such as flies. The smell disappears when the flowers are dried.

Hawthorn Cordial

Hawthorn Cordial

Makes 1 quart

Hawthorn fruit tinctured in alcohol is a potent herbal heart tonic.
It is also a beautiful color and tastes delicious prepared this way. If you start a batch when the haws are ripe in late summer and early fall, it will be ready to enjoy with guests by the winter solstice.

INGREDIENTS

3 cups hawthorn fruit, stems removed

1 pint good-quality brandy (doesn't have to break the bank, but should be something you'd enjoy drinking on its own)

½ cup water

½ cup sugar

1 tablespoon lemon juice

2-inch strip of orange peel, zest only (use a vegetable peeler to remove just the orange part, leaving the bitter white pith behind)

1 cinnamon stick

3 whole allspice

1 whole clove

INSTRUCTIONS

1. Lightly smash each hawthorn fruit by tapping it with a hammer or the bottom of a sturdy bottle. Put the hawthorn into a clean glass jar. Cover it with the brandy. Cover the jar and store it away from direct light at room temperature for 1 month.

2. Put the water, sugar, lemon juice, orange zest, and spices into a small pot. Bring to a boil over high heat, stirring to dissolve the sugar. Turn off the heat. Stir a bit more if the sugar isn't completely dissolved. Cover the pot and let the spiced simple syrup within it cool to room temperature.

3. Strain the hawthorn tincture through a double layer of cheesecloth. Strain the spiced syrup into the hawthorn brandy. Bottle and age for at least 1 month before sampling. Be sure to store the cordial away from direct light or heat. If it's at all cloudy when you're ready to serve it, you can strain it through cheesecloth again, being careful to leave behind any sediment at the bottom of the bottle. Cheers!

Hawthorn Ketchup

Makes 1 pint

Hawthorn ketchup, or haw ketchup as it's sometimes called, is an old-time recipe from the British Isles. It is great used any way that you would use tomato ketchup, although it has its own distinct taste.

INGREDIENTS

2 cups hawthorn fruit, washed and stems removed

½ cup water

⅓ cup apple cider vinegar

¼ cup apple juice

⅓ cup sugar or ¼ cup honey

½ teaspoon salt

½ teaspoon ground allspice

¼ teaspoon freshly ground black pepper

¼ teaspoon ground spicebush berries (optional but really good if you can get them)

INSTRUCTIONS

1. Lightly smash each hawthorn fruit by tapping it with a hammer or the bottom of a sturdy bottle.

2. Put the hawthorn in a pot along with all of the other ingredients and bring to a boil over high heat. Reduce heat and simmer, stirring occasionally, for 25 to 30 minutes until the hawthorn fruits are very soft. Add a little more apple juice if

necessary to prevent the cooking fruit from sticking and burning.

3. Now comes the most tedious part of making hawthorn ketchup: removing the skins and seeds. The easiest way to do this is by putting the cooked mixture through a food mill with the finest-holed blade. But if you don't have a food mill, you can also get away with pushing the mixture first through a colander to break up the fruit, then through a sieve to remove all the seeds. Use the back of a wooden spoon or your clean fingers to push the ketchup through.

4. Return the ketchup to a pot over medium heat and let it simmer until it is as thick as you like it. Remember that hawthorn ketchup tends to thicken even more as it cools, so take it off the heat when it is just slightly runnier than what you want.

5. Transfer the hawthorn ketchup to heatproof jars and cover. Store in the refrigerator for up to 2 months. For longer storage at room temperature, process half-pint canning jars of haw ketchup (leave ½ inch headspace in each jar) in a boiling water bath for 10 minutes. Adjust the canning time if you live at a high altitude (see the Useful Resources section if you're unfamiliar with canning). Once the jars are sealed, your ketchup will keep for at least a year. (It is still safe to eat after that, but the quality will decline.) After you've opened a jar, store it in the refrigerator.

Other fruits you can use in this recipe: apples, crabapples.

Henbit/Deadnettles

LAMIUM AMPLEXICAULE, L. PURPUREUM, AND OTHER EDIBLE *LAMIUMS*

These common weeds grow on several continents and have the virtue of being able to withstand fairly cold temperatures. Henbit is a reliable forage, even during winter in most places.

Find

Henbit likes disturbed soils and often shows up as a garden and farm weed. It is a common plant in city parks.

Identify

All *Lamiums* are in the mint family, and like other members of the Lamiaceae family, they have square stems (roll a stem between your forefinger and thumb and you'll feel the four distinct sides) and opposite leaves (the leaves attach to the stem in aligned pairs).

The leaves of both henbit and other edible, similar-looking plants in the *Lamium* genus (all of which share the unfortunate common name deadnettle) are ½ to 2 inches wide and can be oval-, spade-, or heart-shaped. "Deadnettle" refers to the fact that their leaves look like stinging nettle leaves, but the plant doesn't sting. The leaves have deeply scalloped margins. *L. amplexicaule* leaves attach directly to the stems, and the upper leaf pairs can appear at first glance to be one round leaf surrounding the stem. Other *Lamium* species have short leaf stalks, but the leaf shape is similar. The deep veins

give them an almost quilted appearance. There are hairs on the leaves.

The pink or purple flowers grow in whorls in the leaf axils (where the leaves join the stems). The petals of each small flower are fused into a ½- to ⅔-inch-long tube.

L. purpureum, known by the common name red deadnettle, is a close relative of henbit that is just as winter hardy and widespread, and has similar uses in the kitchen. As its species name suggests, its leaves are tinged with a reddish-purple color. This is especially true at the tops of the plants. The leaves do not clasp the stems the way those of *L. amplexicaule* do.

Henbit and other deadnettles are low-growing plants. The lower stems sprawl on the ground and can root where they touch soil. But the last few inches of the stems usually grow upright.

Sustainably Harvest

Harvesting just the top few inches of the stems of this species in no way threatens the plant's survival. In fact, henbit will grow back even bushier and more tender if you harvest this way.

Eat

Lamium plants may be in the mint family, but they don't taste anything like mint. Rather, they are relatively mild leafy greens that can be eaten raw or cooked.

I think henbit and other *Lamiums* are best when combined with other mild-tasting wild winter greens such as chickweed, or cultivated greens like kale or chard. Henbit holds up well to strong seasoning: Garlic and/or ginger are good choices depending on the direction your recipe is taking. Or try the mushroom-henbit match in this pasta recipe.

Henbit Noodles with Creamy Wild Mushroom Sauce

Serves 4

The trouble with henbit is that its flavor includes musty overtones. Someone commented on a Facebook wild edibles group that henbit has a "mushroomy taste." Bingo! Pairing henbit with wild mushrooms turns that characteristic into a pro rather than a con.

NOODLES INGREDIENTS

½ pound henbit leaves

1 garlic clove, peeled

2 eggs

1 tablespoon extra-virgin olive oil

1½ cups all-purpose flour, divided

¾ cup semolina flour

SAUCE INGREDIENTS

1 cup fresh mushrooms or reconstituted from dried (save soaking liquid), chopped

2 tablespoons butter

Salt to taste

1 cup mushroom stock (or soaking liquid from dried mushrooms) or vegetable or chicken stock

¼ teaspoon dried or ¾ teaspoon fresh thyme leaves

½ cup light cream

¾ cup Parmesan or Pecorino Romano cheese, freshly grated and divided

Freshly ground black pepper

FOR THE NOODLES

1. Put the henbit leaves into a pot with ¼ cup water. Cook them over medium heat until the leaves are completely wilted, about 5 minutes. Drain in a colander and immediately run cold water over them. Squeeze out as much liquid as possible.

2. Pulse the cooked henbit and the peeled garlic in a food processor (or finely mince both with a knife, or put the garlic through a garlic press and chop the henbit separately).

3. Add the eggs and olive oil and purée the ingredients.

4. Reserve ½ cup of the all-purpose flour. Whisk together the rest of the all-purpose flour and the

semolina flour in a large bowl. Dump the contents of the bowl out onto a clean counter or cutting board. Make a well in the center.

5. Pour the egg-henbit mixture into the well in the center of the flour. Mix the flour into the liquid mixture with a fork.

6. Knead the mixture by hand for 10 minutes (or in a stand mixer with the bread hook attachment, or in a food processor with the dough blade) until the dough comes together into a ball. Kneading by hand is the best option because you have more control over how much flour ends up in the dough. Stop incorporating more as soon as you can knead the dough without it sticking to your fingers.

7. Cover the dough with a clean, damp kitchen towel and let it rest for 30 minutes.

8. Lightly dust your work surface. Cut the rested dough into quarters. Roll one of the quarters out with a rolling pin or an empty wine bottle until it is as thin as you can get it. Turn the dough over frequently while you roll it out and dust with additional flour as necessary to prevent it from sticking to your rolling implement.

9. Give the rolled-out dough one more light sprinkling of flour, then roll it up loosely. Cut crosswise so that it forms coils of ¼- to ½-inch-wide noodles. Uncoil the coils and dust them with additional flour.

FOR THE SAUCE

1. If you're using dried mushrooms, first soak them in boiling-hot water for 15 minutes. Drain (reserve the soaking liquid) and squeeze out as much liquid as possible. Whether you started out with fresh mushrooms or dried, coarsely chop them before the next step.

continued

2. Melt the butter in a skillet over medium-low heat. Add the mushrooms and a little salt. Cook, stirring often, until the mushrooms give up their liquid and then most of the liquid evaporates.

3. Add 2 tablespoons all-purpose flour and cook, stirring, for 2 minutes. Raise the heat to medium high.

4. Add the mushroom soaking liquid (if you started with dried mushrooms) and/or the stock a small splash at a time, stirring constantly. Add the thyme with the first addition of liquid. Allow the sauce to thicken after each addition of liquid before you add more. When it is all the consistency of a thick gravy, turn off the heat and stir in the cream and salt and pepper to taste.

PUTTING IT ALL TOGETHER

1. Bring a large pot of salted water to a boil. Add the fresh henbit noodles and stir gently. Cook for 3 minutes. Drain. Return to the pot and add the sauce and ¼ cup of the grated cheese. Toss gently to coat the noodles with the sauce. Add salt and pepper to taste.

2. Serve with additional grated cheese and a little minced fresh henbit or parsley sprinkled over as a garnish.

Other plants you can use in this recipe: pretty much any leafy green will work, but for vibrant color plus extra nutritional value, go for nettles.

Japanese Knotweed
POLYGONUM CUSPIDATUM
SYN. *FALLOPIA JAPONICA*

Edible both raw and cooked, and equally good in both savory and sweet recipes, Japanese knotweed's tangy stalks taste like rhubarb with grassy flavor notes.

Find

Japanese knotweed usually grows in disturbed and rocky soils. It thrives in both partial shade and full sun. It is especially prolific in urban and suburban areas such as gardens and city parks, but also alongside country roads.

Collect Japanese knotweed shoots in early spring, when they are thick and unbranched.

Identify

When Japanese knotweed first emerges in spring, its hollow shoots pop straight up with just a few leaves, including those still tightly furled at the tips. The skin of the shoots as well as the older stalks is pale green mottled with reddish spots. The shoots, and later the mature branching stalks, are jointed and have prominent, swollen nodes where the leaves emerge in older plants.

Those jointed stalks lead some people to think that Japanese knotweed is a type of bamboo. But *Polygonum cuspidatum* never becomes woody the way bamboo does, and its leaves are a completely different shape. Bamboo's leaves are slender and lance-shaped, whereas Japanese knotweed leaves are shaped like a gardener's trowel. They are almost as wide as they are long, pointed on one end, and flat or barely curved across the base end. The leaves can grow to be 4 to 6 inches long.

If you weren't prejudiced by the knowledge that this is a highly invasive weed, you might notice that Japanese knotweed is really quite attractive with its jointed, bamboo-like, reddish stalks and plumes of off-white flowers. In fact, Japanese knotweed was originally introduced to North America as an attractive ornamental landscaping plant.

The seeds are in triangular achenes about ⅛ inch across, tan to brown, and glossy.

The roots are impressive, with clumps up to 2 feet in diameter, woody, and convoluted. They have challenged many a gardener's attempts to eradicate them (including this gardener).

Sustainably Harvest

Japanese knotweed is aggressively invasive. It will regrow if even an inch of one of its massive root systems is left in the ground, as well as spread by wind-blown seeds. Harvest as much of this plant as you can possibly use . . . please!

Harvest by cutting young, fat shoots off a couple of inches above the ground. If the plants are more than a foot high but still unbranched, harvest by trying to snap them off by hand. As with asparagus, the point at which they snap off easily with a clean break marks the separation between the tender, edible upper part of the shoot and the too-tough-to-eat lower portion.

Eat

A versatile ingredient, Japanese knotweed's early spring stalks can be eaten raw or cooked in savory as well as sweet dishes. "Use like rhubarb" is the usual culinary advice with this plant because it has a similar sourness and, like rhubarb, transforms from crunchy to fall-apart soft when cooked. But that is where the similarity ends. Japanese knotweed's flavor includes grassy notes rhubarb lacks, which can be quite pleasant, depending on how used.

The leaves are also edible, although not particularly interesting in my opinion.

If the skins are so thin that they are almost impossible to peel off in strips, don't bother peeling. But if it is easy to peel the skins off the stalks in strips, then do it—the texture will be much more appealing, both raw and cooked.

Older Japanese knotweed stalks are still edible, it's just that they get too fibrous to use in recipes. However, they're refreshing as a lemony chew while in the field.

Hummus-Stuffed Knotweed Snacks

Serves 8 to 10

This recipe reminds me of the cream cheese–stuffed celery served during the 1970s, only better. The crunchy-plus-creamy combination is similar, but the tangy, complex flavors of the raw knotweed and the lemony hummus are more interesting. It makes an easy, conversation-starting party hors d'oeuvre.

INGREDIENTS

½ pound Japanese knotweed stalks, washed and leaves removed

1 cup hummus (you can use a high-quality store-bought brand, but I've included a good, easy hummus recipe on the following page)

INSTRUCTIONS

1. Peel the knotweed and cut the stalks crosswise into 1-inch lengths. Discard the solid sections at the leaf nodes, and only use the hollow pieces.

2. Use a table knife to spread some of the hummus into one side of a knotweed tube. Turn the knotweed over and stuff the other side with hummus (this is why the pieces are only 1 inch long—any longer and it is a real pain to try to get the hummus all the way into the center of each tube). Repeat with the remaining knotweed tubes.

Easy Homemade Hummus

Makes 2 cups

INGREDIENTS

2 cups cooked chickpeas (garbanzo beans), drained, but reserve some of the cooking liquid (yeah, that includes the goop in the can if you are taking some help from the store)

½ cup tahini

¼ cup extra-virgin olive oil

2 teaspoons garlic, minced (wild garlic is excellent here)

2 teaspoons ground cumin

Juice of 1 to 2 lemons (depending on how big and juicy they are)

Salt

INSTRUCTIONS

1. Purée the chickpeas, tahini, oil, garlic, cumin, and the juice of 1 lemon in a food processor. Add some of the cooking liquid or some water a little at a time until you have a smooth purée. You may need to stop and scrape down the sides of the food processor vessel once or twice during the process.

2. Taste the hummus, and mix in salt and additional lemon juice to taste, if needed.

Sweet and Tangy Knotweed Bars

Makes 16 bars

All of the tanginess in these yummy bars comes from the Japanese knotweed. Frozen knotweed works just as well as fresh here; just measure it frozen and then let it thaw completely before proceeding with the recipe.

INGREDIENTS

1½ cups Japanese knotweed stalks, peeled and finely chopped

¾ cup brown sugar, divided

4 tablespoons water, divided

4 teaspoons cornstarch

1 cup rolled oats

½ cup all-purpose flour

½ teaspoon salt

5 tablespoons butter, melted

INSTRUCTIONS

1. Preheat the oven to 350°F. Lightly grease an 8-inch square baking pan with butter or coconut oil.

2. Combine the chopped knotweed stalks, ½ cup of the brown sugar, and 3 tablespoons of the water in a medium pot. Bring to a boil over high heat, then reduce the heat and simmer, stirring often, until the knotweed softens and starts to fall apart.

3. Stir the cornstarch and remaining tablespoon of water together until you have a smooth paste. Stir the cornstarch into the knotweed mixture. Raise the heat to high and cook, stirring constantly, until it thickens. Remove from heat and set aside.

4. In a large bowl, stir together the oats, flour, salt, and remaining ¼ cup of brown sugar. Add the butter and stir until you've got a crumbly but well-combined mixture.

continued

5. Press half of the oat mixture into the baking pan. Spread all of the knotweed filling over the top. Top that with the rest of the oat mixture.

6. Bake for 25 to 30 minutes, until the top is golden brown. Cool completely on a rack, then cut into bars.

Jerusalem Artichoke/Sunchoke
HELIANTHUS TUBEROSUS

Jerusalem artichokes are neither artichokes nor from Jerusalem. The tubers of this native North American sunflower relative are good both raw and cooked.

Find

Look for Jerusalem artichokes—J-chokes for short—in full sun sites near rivers, roads, and fence lines. Although I sometimes find them growing in partial sun, the tubers are bigger and more plentiful where the plants get ample light. They often form large colonies.

You'll find the tubers tastiest in late fall, winter, and early spring. To locate your digging spot during those months, look for the tall stalks (up to 10 feet tall, though often several feet shorter) with flower and seedhead remains and alternately arranged leaf scars or leaf remnants on the upper parts of the stalks.

Identify

During the growing months, the tall Jerusalem artichoke stalks have ovate leaves (roughly oval with pointed tips). The lower leaves grow in oppositely arranged pairs, but the smaller leaves higher up the stalks may be alternately arranged. The leaves have winged leaf stalks (petioles), which appear as two thin strips running from the bottom of the leaf down either side of the stalk. Each leaf has three

prominent veins. The two outer veins run about half-way down the leaf, while the midvein runs from the leaf base to the leaf tip. The leaves are covered with short, stiff hairs that give them a sandpapery feel.

The stems often have a purplish-red tint. Like the leaves, the stems have a sandpapery texture thanks to the coarse hairs that cover them.

The 3-inch-diameter yellow flowers bloom from late summer into mid-autumn. They look like small versions of their cousin, the sunflower, but without the dark brown center. They have a yellow to yellowish-orange central disk instead, which is about an inch in diameter. The 10 to 20 yellow petals circling the disk are actually ray flowers.

The wild tubers are thin skinned and finger shaped. The skin is sometimes tinged with purple or red, but other times it is beige. Garden-grown tubers, by contrast, are more plentiful per plant. Their skin is whitish and they are shaped like convoluted, lumpy potatoes.

Sustainably Harvest

There's really not much of a sustainability issue with sunchokes. But if you want to be really, really sure the plants will regrow next year in the patch you just harvested from, simply replant a few of the smaller tubers that you dig up. Even a chip off one of the tubers can be enough to regrow a new plant.

To harvest Jerusalem artichoke tubers, it helps to identify a patch of the plants in late summer or early fall, when they are in bloom. Mark that loca-tion and come back after a few frosts to harvest. The gnarly tubers are good cooked or raw, especially after they have been in the ground for a freezing cold night or two. In fact, I dislike the taste of them until after a couple of freezes have converted some of their musty-flavored starch into sugar. After that, I find them delicious.

The plants are often prolific tuber producers, but you might have to dig quite far down to get to them. Grab a shovel and dig about 1 to 1½ feet deep around the stalks and you will unearth the tubers.

Eat

As mentioned earlier, the tubers are much better once they've been through a few frosts. Before that, they can have a funky off taste. That off taste comes from a starch called inulin. Cold weather converts inulin into fructose, resulting in a sweeter flavor.

Once they've been zapped by the cold, enjoy Jerusalem artichokes raw in salads, where they have a crunchy texture something like jicama. Steamed, boiled, or roasted, the texture is more like potato. The taste of cooked Jerusalem artichokes pairs well with earthy flavors such as mushrooms.

Jerusalem artichokes are terrific pickled. A bonus is that the pickling process seems to help minimize the infamous fart factor associated with this vegetable, although that might have more to do with the smaller quantity you're likely to eat of the pickled version compared to other preparations.

Sweet and Sour Pickled J-Chokes

Makes 3 pints

Crisp, sweet, spicy, and sour, these pickled Jerusalem artichoke tubers are a tasty snack on their own. They are also excellent served alongside curried dishes.

BRINING

1. In a large bowl, combine:
    Juice of 1 lemon or 2 tablespoons vinegar
    2 cups water

2. Scrub clean (no need to peel):
    1½ pounds Jerusalem artichokes

3. Drop the chopped J-choke pieces into the acidulated water so that they do not discolor.
    In a second container, combine:
    3 cups water
    2 tablespoons kosher or other non-iodized salt

4. Stir to dissolve the salt. Drain the J-chokes that are in the acidulated water and transfer them to the salt brine. Cover and refrigerate for 8 hours (or

continued

as long as 24—I did this accidentally the second time I made these and didn't notice any difference in the finished pickle).

PICKLING

5. Get your boiling water bath set up if you plan on canning your jars of pickled Jerusalem artichokes (see Canning below).

   Combine the following ingredients in a small saucepan over high heat and bring them to a boil. Reduce the heat and simmer for 5 minutes:

   1 cup cider vinegar

   ½ cup rice vinegar

   ½ cup water

   ¼ cup sugar

   1 tablespoon turmeric powder

   1 teaspoon mustard seeds

   ½ teaspoon black peppercorns

   ½ teaspoon whole allspice or spicebush berries
      (leave out the black pepper if using spicebush)

   ¼ teaspoon coriander seeds

   ¼ teaspoon cumin seeds

   1 whole clove

FILLING THE JARS

6. Drain the J-chokes and rinse them under cold water to remove excess salt. Load them into clean jars, leaving an inch of headspace.

   Divide the following among the jars:

   1 bay leaf, broken into a few pieces

   3 small hot chili peppers (optional)

7. Pour the hot brine over the J-chokes, bay leaf, and chilies, leaving at least ½ inch of headspace (the liquid should completely cover the other ingredients).

   It's fine if some of the spices from the brine end up in the jars, but if you want a clear, golden brine rather than turmeric "mud" at the bottom of the jars, first strain it through a muslin bag or several layers of cheesecloth.

   Wipe the rims of the jars dry and secure the lids.

CANNING (OPTIONAL)

If you'd rather skip the canning step, store your pickled Jerusalem artichokes in the refrigerator.

If you want to seal pickles in the jars and safely store them at room temperature, process them in a boiling water bath for 5 minutes.

Jerusalem Artichoke and Roasted Garlic Soup with Mushrooms

Serves 6

The earthiness of Jerusalem artichokes goes beautifully with sweet parsnip roots, roasted garlic, and the deep mushroom flavor of shiitake or maitake (hen of the woods) mushrooms.

INGREDIENTS

1 small head of garlic (about 8 cloves)

1½ tablespoons olive oil, divided

2 pounds Jerusalem artichokes, scrubbed clean and chopped into 1-inch pieces

2 large parsnips, peeled and chopped into 1-inch pieces

½ cup celery, chopped

¼ pound shiitake or maitake mushrooms, thinly sliced

2 large leeks, white and tender green parts, sliced

2 teaspoons fresh or ¾ teaspoon dried thyme leaves

½ teaspoon freshly ground black pepper

¼ cup light cream (optional)

Salt

continued

INSTRUCTIONS

1. Slice the top off the head of garlic cloves and baste the cut-open surface with 1 teaspoon of the olive oil. Loosely wrap the head of garlic in aluminum foil and roast in a 400°F oven for 30 to 40 minutes until mushy-soft when squeezed, but not burnt.

2. While the garlic is roasting, put the Jerusalem artichokes, parsnips, and celery into a large pot. Add 1 quart water and bring to a boil over high heat. Reduce the heat and simmer, partially covered, until the vegetables are tender.

3. Heat 2 teaspoons of the oil in a large skillet. Add the mushrooms and cook over medium-low heat, stirring often, until the liquid in the mushrooms is first released and then is reabsorbed and/or evaporated. Remove the mushrooms from the pan and set aside.

4. In the same skillet, heat the remaining 1½ teaspoons of oil. Add the leeks and cook, stirring frequently, until they soften.

5. Add the leeks to the Jerusalem artichokes, parsnips, and celery. Purée with an immersion blender, or in two batches in a regular blender. Return to the pot. Add the thyme and black pepper and more water if the soup seems too thick. Simmer, partially covered, for 10 minutes.

6. Remove from the heat and stir in the cream and salt to taste. Divide the mushrooms between six bowls and ladle the soup over them. Serve hot.

Juneberry/Serviceberry
AMELANCHIER SPECIES

One of the first fruits to ripen each year, juneberries are juicy and enjoyable raw . . . but save some for baking because they also make fantastic pie!

Find
Amelanchier species grow as tall shrubs or small trees in full to partial sunlight. They grow wild in open woodlands and beside lakes, but are also frequently planted by landscapers in city parks.

Identify
Ripe juneberries look like blueberries growing on a tree—they even have the five-pointed crown on one end that blueberries have. As the berries ripen they turn from green to red and eventually dark purple.

A couple of months before the berries appear, the flowers put on quite a show. They have five strap-like white petals and numerous stamens at their centers. *Amelanchier* flowers bloom before any leaves emerge each spring, and are often still hanging on even once the alternate, oval leaves start to unfurl. The leaves have fine teeth along their edges, thin leafstalks, and turn a beautiful golden-amber color in the fall.

Juneberries have gray bark that is usually smooth but sometimes develops shallow vertical grooves as the plants mature.

Sustainably Harvest
Picking juneberry fruit in no way harms the parent tree. The berries don't all ripen at the same time, so expect the harvest to last for 2 or 3 weeks. Juneberries produce one of the first fruits of the year, ripening sometime between late spring and early summer. They are temperature triggered out of winter dormancy, so a warm winter yields an earlier spring crop, whereas a cold spring can delay flowering and fruiting.

Eat
Juneberries are delicious raw anytime from when they first turn red all the way through their fully ripe, deep purple stage. They freeze well and are delicious in pies, jam, and pancakes. Their juicy pulp is mild and sweet, but it's the tiny seeds that complete juneberry's flavor: they have a light almond taste.

Juneberry Pie

Makes one 8-inch pie

This is one of my all-time favorite pies! You can use fresh or frozen juneberries in this recipe. If you use frozen ones, measure them while they are still frozen, but then thaw them before proceeding with the recipe.

CRUST INGREDIENTS

2½ cups flour

1 teaspoon salt

1 cup (2 sticks) cold butter cut into small pieces

¼ cup lard or coconut oil

6 to 7 tablespoons ice water

INSTRUCTIONS

1. Whisk the flour and salt together. Work in the butter and lard a little at a time until the texture is somewhere between coarse cornmeal and small peas. You can do this and the next step either with a pastry blender, a fork, or by pulsing in a food processor.

2. Gradually mix the water into the other ingredients. The dough's final texture should be loose, but it should hold together when you pinch a bit of it.

3. Divide the dough in half and shape each half into a ball. Flatten the balls into disks, then roll them out until they are thin enough to cover the bottom of an 8-inch pie pan with some overhang. Line the pan with one of the crusts. Save the other for the top. If you want to make a lattice top crust, slice the second crust into ½-inch-wide strips. It is not necessary to prebake the bottom crust.

FILLING INGREDIENTS

4½ cups fresh or frozen juneberries

¾ cup sugar

3½ tablespoons cornstarch

1 tablespoon lemon juice

½ teaspoon ground spicebush berries or ¼ teaspoon
    ground allspice plus ⅛ teaspoon ground pepper

⅛ teaspoon salt

1 tablespoon unsalted butter, cut into small pieces

1. In a large bowl, combine the juneberries, sugar,
   cornstarch, lemon juice, spices, and salt. Let sit
   for 15 minutes while you preheat the oven to
   425°F. You'll use the butter in the next step.

continued

ASSEMBLE AND BAKE

1. Spoon the filling into the crust-lined 8-inch pie pan. Dot the filling with the butter.

2. Cover with the second crust or create a crisscross lattice. Fold the overhang of the bottom crust over the edges of the top crust. Trim off any excess and crimp with your fingers or press with the tines of a fork to seal the edges of the crusts.

3. Bake for 30 minutes. Take the pie out of the oven and loosely wrap the outer edge of the crust in aluminum foil to prevent it from burning. Return the pie to the oven and bake it for another 30 minutes.

4. Remove the pie from the oven and let it cool on a rack for at least 45 minutes before slicing and serving.

Lamb's Quarters/Wild Spinach/ Goosefoot

CHENOPODIUM ALBUM, C. MURALE, AND OTHER EDIBLE CHENOPODIUM SPECIES

It is increasingly common to find this delicious plant for sale at farmers markets and co-ops, but it grows wild on several continents and you can likely collect it for free where you live. When cooked, it has the same silky texture as spinach and a similar taste.

Find
A common weed in gardens and parks and on farms, look for lamb's quarters in full to partial sunlight near human habitats.

Identify
One of the key identifiers for lamb's quarters (*Chenopodium album*, but not *Chenopodium murales*) is the mealy white coating that you can rub off the leaves. It is especially obvious on the younger leaves. Other plants with similarly shaped leaves lack this waxy coating. Also, unlike some of its botanical kin, lamb's quarters species have no discernible scent.

Lamb's quarters leaves join the stem in an alternate arrangement with the exception of the first two true leaves. The leaves are roughly triangular with soft, rounded teeth along the margins, but leaves near the top of older plants are usually smaller and elliptical.

The stems are frequently tinged with red or purple and grooved; in some horticultural varieties even the leaves are bright magenta. The branching

clusters of tiny green flowers are not particularly eye-catching, and the seeds, when ripe, are small and black within those clusters.

Sustainably Harvest
Lamb's quarters is considered invasive in some parts of the world. You are not endangering this plant by gathering and eating it. However, if you want to collect the seeds as well as the leaves, leave a few of the plants unharvested so that they can grow to maturity.

When the plants are young the stems will snap off easily. At that stage, harvest and eat both the stems and leaves. As the plants mature, the stems become too tough to be good eating, but the leaves remain nonbitter and tasty. At that stage, pinch off the clusters of leaves that emerge at the axils, leaving the tough stems behind.

Chenopodium species are prone to leaf miner bugs. These tiny insects tunnel between the upper and lower epidermis of leaves, leaving clear to white winding trails where they have fed. Usually these infestations don't show up until midsummer, so there's plenty of opportunity to gather the leaves and tender stems in midspring through early summer while they still look healthy. However, if you do find plants that are uninfested later in the growing season, the leaves remain tender and not at all bitter, even after the plants flower and go to seed.

To harvest the seeds, wait until they are mature enough to shake out into the palm of your hand when you jostle a seedhead. You can either shake the seedheads into a bag or, more efficiently, snip whole ripe seedheads off into a cloth or paper bag. Let them dry for a few days until many of the seeds have fallen out. Rub the seedheads lightly to release the remaining seeds. Winnow the chaff from the seeds by tossing in a large bowl in front of a fan or in the presence of a light breeze—the seeds will drop back into the bowl while the chaff blows away.

Eat

Lamb's quarters leaves and young stems deserve their other common name, wild spinach: Their taste is very similar, as is their texture once cooked. Usually I consider the frequently repeated instruction to use any wild leafy green "like spinach" to be a cop-out, but with *Chenopodium* species it is actually true.

The seeds can be cooked using a ratio of two parts water to one part seeds; the result is something like quinoa. Or add them dry to crackers, muffins, and breads.

Wild Colcannon

Serves 6

The Irish side of me insists that I share this recipe. Because really, what's not to love? Creamy mashed potatoes made lively with a bit of onion and bright green lamb's quarters leaves replacing the usual kale or cabbage. I know this is supposed to be a side dish, but I confess it's usually the main course on my plate—I love it that much.

INGREDIENTS

2 pounds Yukon gold, russet, or other prime mashable
potatoes (they should be a floury type rather than
waxy)

⅓ cup butter

¼ cup milk

Couple scrapes of freshly ground nutmeg

4 ounces bacon, chopped into very small pieces
(optional) or 1 tablespoon vegetable oil

½ cup chopped scallions, white and tender green parts

1 pound lamb's quarters leaves and tender stems, washed
and finely chopped

Salt

continued

INSTRUCTIONS

1. Peel the potatoes and cut them into large chunks. Put into a pot and cover with cold water. Bring to a boil over high heat and cook until they are soft enough to fall apart when pricked with a fork. Drain the potatoes in a colander and then return them to the pot.

2. Add the butter, milk, and nutmeg, and mash while the potatoes are still hot. Do *not* use an immersion blender or you will end up with glue rather than mashed potatoes.

3. Cook the bacon, if using, over medium heat in a frying pan until the meat is starting to get crispy and most of the fat has rendered out.

Alternatively, heat the oil in a frying pan until it shimmers, but don't let it smoke. Add the scallions and cook for 1 minute. Add the lamb's quarters and cook, stirring, until the greens are completely wilted and the scallions are tender, about 5 minutes.

4. Add the greens and scallions to the potatoes and mash together. Add salt to taste. Serve hot. (Hint: leftover colcannon makes THE best topping for shepherd's pie.)

Any mild-flavored leafy green will work in colcannon, but next best after lamb's quarters are amaranth, quickweed, and nettles.

Wild Spinach Dip in a Bread Bowl

Makes 1 pint

The old-school version of spinach dip in a bread bowl involved dumping together powdered soup mix, frozen spinach, mayo, and sour cream. This is that dip, but with wild spinach (a.k.a. lamb's quarters), dehydrated onion, and wild garlic powder instead of the soup mix, though you could use store-bought garlic powder if need be. Don't think you'll get a better result by using raw or freshly sautéed onion and garlic: the dehydrated versions have a distinct taste that encompasses the best of what that powdered soup mix used to do for this classic dip.

INGREDIENTS

1½ cups cooked, chopped lamb's quarters (from about 1 pound of fresh leaves and tender stems)

1 cup sour cream

1 cup mayonnaise

¼ cup Parmesan or Romano cheese, grated

1 teaspoon wild (or not so wild) garlic powder

1½ teaspoons dried onions or 1 teaspoon dried onion powder

½ teaspoon freshly ground black pepper

Salt to taste (you may not need any if you used Romano for the cheese)

1 small round loaf of bread (sourdough recommended)

INSTRUCTIONS

1. Make sure to squeeze as much liquid out of the cooked greens as possible. In a bowl, combine the wild spinach with all the other ingredients except for the bread. Cover, refrigerate, and wait at least an hour for the flavors to develop before serving.

continued

2. To serve in true '70s retro fashion, slice off the top of the loaf of bread. Tear out the inside of the main portion of the loaf to create a bowl (I save the torn-out bread to make bread crumbs). Scoop the wild spinach dip into the bread bowl and serve with crackers or crudités.

Other plants that are great in this recipe: You could use pretty much any leafy green, but amaranth leaves come closest in flavor. To go in a spicier direction, try using mustard or garlic mustard greens.

Maitake/Hen of the Woods
GRIFOLA FRONDOSA

When a forager says, "I found hen today," he isn't talking about poultry. He is talking about one of the best and easiest to identify edible wild mushrooms: hen of the woods, also known as maitake.

Find

Grifola frondosa grows in large clumps at the base of trees (especially oak trees) and on tree stumps. It has a firm, fleshy but tender texture. Maitake sometimes appears to be growing on lawn or bare ground, but if you look up or dig around you'll realize it is growing on buried wood or tree roots.

Identify

When I said large clumps, I meant *large*, sometimes measuring as much as 2 feet across. The clumps look like big corsages made up of gray, brown, and/or off-white ruffles. The layers have pores on their cream-colored undersurface rather than gills.

Beginners might confuse turkey tail mushroom (*Trametes versicolor*) with maitake. Turkey tail is much smaller and usually spread out along a log rather than in a rounded clump at the base of a tree trunk. It is thin, tough, and leathery or papery, rather than tender and meaty like maitake. Although it has medicinal properties and sometimes the flavor is interesting enough for soup stocks and infusions, the texture is pretty much shoe leather (unlike the firm but tender texture of hen of the woods).

Sustainably Harvest

As with all mushrooms, the aboveground fruiting body that we eat is only a fraction of the actual mushroom. The mycelium, the "roots" of the mushroom, are invisible to us when we look at a massive clump of hen. You are not harvesting the mycelium when you cut off the aboveground parts of this mushroom. Use a sharp knife and cut near the base. Have a sturdy, breathable container—a basket, cloth, or paper bag is best—ready to hold your harvest.

Eat

The only tedious part of working with maitake is scrubbing away the dirt that clings between the ruffled layers along with any insects that might have moved in. I like to slice the mushroom into pieces about ⅓-inch thick before cleaning them with a vegetable brush. Go ahead and do this under running water or in a large bowl of water—it will save you much trouble. The culinary advice about not washing mushrooms in water doesn't apply here.

Hen of the woods needs to be cooked before you eat it, as do all wild mushrooms (and some would say the cultivated ones, too). Sautéed in the lipid of your choice (butter, olive oil, duck fat, etc.), it is sublime.

Raw hen dries and freezes beautifully and needs no preparation for freezing other than the thorough cleaning mentioned above.

To dry hen of the woods, place slices of it in a dehydrator at 125°F or in your oven on the lowest setting with the door propped open with a dish towel or wooden spoon. The mushrooms should be crispy-dry within 4 hours.

To use dried hen of the woods, first reconstitute it by pouring boiling water over the dried mushrooms and letting them sit in the hot water for 15 minutes. Remove the mushrooms. Carefully pour off and save the soaking liquid, leaving behind the last bit of liquid with any gritty sediment. Use the mushrooms in any mushroom recipe. Use the soaking liquid in soups, sauces, or (heaven!) risotto.

To freeze hen of the woods, clean, chop, and spread it in a single layer on a baking sheet. Freeze, uncovered, for 1 to 2 hours. Transfer to freezer bags or containers. This method keeps the pieces of mushroom loose so that later on you can take out just what you need for a recipe. If you just put raw chopped hen of the woods into a container or bag and freeze it, you end up with a big brick of chopped mushroom, probably far more than you need for any one recipe.

Hen Sauce

Makes 4 to 6 servings

How can something so simple be so good? Serve this over pasta, polenta, rice, poultry, meat, or roasted root vegetables for a delicious main course. You can use frozen or reconstituted dried hen of the woods in this recipe when you don't have fresh on hand.

INGREDIENTS

2 tablespoons butter

1 tablespoon olive oil

3 cups hen of the woods mushrooms, cleaned and slivered

¼ teaspoon sea salt

1 teaspoon garlic, minced or pressed

¼ cup dry white wine

Additional salt plus freshly ground black pepper to taste

INSTRUCTIONS

1. Heat the butter and oil in a skillet over low heat.

2. Add the mushrooms and ¼ teaspoon salt and cook, stirring occasionally, until the mushrooms are tender, 10 to 15 minutes.

3. Add the garlic and cook, stirring, for 1 more minute.

4. Raise the heat to medium-high. Add the wine and cook, stirring, until most of the wine has evaporated or been absorbed by the mushrooms. Add salt and pepper to taste.

Most edible wild mushrooms will work in this recipe, but chicken of the woods and oyster mushroom are especially good.

Mallow

MALVA NEGLECTA, M. SYLVESTRIS, M. PARVIFLORA, AND OTHER EDIBLE *MALVA* SPECIES

Sold as a prized vegetable in African and Middle Eastern markets, mallow is usually overlooked as a common weed in North America and Europe. That's a shame because this plant boasts three delicious edible parts, one of which you can even use as a meringue-like dessert topping! It is also a potent herbal medicine.

Find

Mallow plants prefer full to partial sunlight and the disturbed soils that come with human habitation. They are frequently found near roadsides, fallow fields, and urban parks and gardens.

Identify

Depending on the species of mallow that you find, the plant may be anything from a mat-like, low-growing species to one that is 3 or 4 feet tall. What all of the mallow species have in common are round or faintly lobed hairy leaves that have distinctive pleats in them, like the folds of a fan. Mallow leaves remind many people of geranium (*Pelargonium*) leaves, but geranium leaves don't have those obvious creases. Mallow leaves grow alternately on the stems and have fairly long stalks.

The flowers have five petals that are usually pink or lavender. They emerge from the points where

the leaves join the stems and look something like miniature hollyhock flowers.

The fruits (seedheads) are sometimes called "cheeses" or "wheels," and they do indeed look like tiny, flattened wheels of cheese divided into wedge-shaped segments. The whole wheel is incompletely wrapped in a five-pointed sheath.

Sustainably Harvest

Mallow leaves are at their best when they first unfurl. Unlike many other wild leafy greens, mallow leaves do not become bitter after the plant starts flowering, and in theory you could still harvest them then. However, as the plants mature their leaves are prone to a rust (fungal disease) that appears as unsightly, orangish bumps on the leaves. Not what I want for dinner, so I pick the full-size but young leaves before the rust develops.

The flowers can be picked at any time, but leave plenty on the plants to develop into the "cheeses."

Go ahead and pick those "cheeses" while they're still green, but leave some to ripen and seed the next generation of plants.

Eat

Mallow leaves are too fuzzy to eat raw, at least to my taste. And cooked, they develop a mucilaginous texture that isn't necessarily pleasant. The key is to use them either in recipes like soup, where that slimy texture becomes a useful thickening agent, or in small amounts as in the stuffed mallows recipe below. You can also dry the leaves and use them to make an infusion that is beneficial for respiratory complaints like nagging coughs, as well as for upset stomachs.

The flowers are mild and mostly worth using as an attractive salad garnish.

The immature seedheads are where things really get interesting. You can eat them raw or cook them as a vegetable, but their special use is as an egg white–like ingredient. You can use them to make vegan mayonnaise or whip them into a meringue-like foam. You can even turn them into marshmallows (or mallow mallows, as a colleague calls them). You also can pickle the green seedheads and use them like capers.

Stuffed Mallows

Makes 15 to 20 rolls

Similar to dolmades (Greek stuffed grape leaves), stuffed mallow leaves are quicker to make because the leaves are more delicate than grapevine leaves: You don't need to blanch them first, and the cooking time is about half as long. The mucilaginous stuff that comes out of the leaves isn't at all slimy here, but rather works to thicken the sauce the rolls cook in.

2 tablespoons finely chopped Moroccan-style preserved lemon or juice and zest of 1 lemon

1 tablespoon finely chopped fresh mint or 1 teaspoon dried and crushed mint

1 teaspoon salt

½ teaspoon freshly ground black pepper

½ teaspoon ground cumin

15 to 20 large mallow leaves

Juice of 2 lemons

¾ cup extra-virgin olive oil

¾ cup water

1 tablespoon honey

1 lemon, cut into 6 wedges

INGREDIENTS

1 cup long-grain brown or basmati rice

2 cups water

3 large tomatoes

⅓ cup onion, finely chopped

¼ cup pine nuts

¼ cup walnuts, crushed

¼ cup fresh parsley, minced

6 garlic cloves, peeled and divided

INSTRUCTIONS

1. Put the rice and water in a small pot with a lid. Bring to a boil, then reduce the heat and simmer, covered, until the rice is cooked. Drain off any excess water.

2. While the rice is cooking, finely chop one of the tomatoes. Finely chop two of the garlic cloves. In a large bowl, combine the tomatoes and garlic

the leaf away from you until you have something that looks like a very short cigar. Place the stuffed mallow leaf in the pot on top of the tomato slices with the side of the roll that shows the edge of the leaf facing down (so that it doesn't unroll). Repeat with the remaining mallow leaves, snuggling each roll next to one of the others, always "seam side" down. Continue until the whole bottom of the pot is covered with the tightly packed-in mallow rolls. (If you don't have enough to cover the entire bottom of the pot, you need a smaller pot.)

5.   Tuck the remaining garlic cloves in between the stuffed mallow rolls.

6.   In a small bowl, whisk together the lemon juice, olive oil, water, and honey. Pour the liquid over the stuffed mallows. Place a plate that will fit inside the pot on top of the mallow rolls to keep them from floating while they cook. Cover the pot with its lid and let the rolls cook over low heat for 45 minutes. Check on them occasionally and, if necessary, add a tiny bit of water to keep them from drying out (this is almost never needed).

7.   Let the mallow rolls cool to room temperature before removing them from the pot. Serve stuffed mallows with some of the tomatoes from the pot and lemon wedges. They will keep, refrigerated, for several days, but be sure to let them come back to room temperature before serving.

with the onion, nuts, parsley, half of the garlic, lemon, mint, and seasonings.

3.   Slice the remaining tomatoes into rounds ¼- to ½-inch thick, then line the bottom of a large pot with them.

4.   Snip off any stems from the mallow leaves. Place a leaf on your work surface with the shinier side down and the stem end toward you. Put about a teaspoon of the stuffing mixture onto the leaf just above where the stem was attached. Close the sides of the leaf over the stuffing, then roll

Mallow Foam

Makes 1 to 2 cups foam

INGREDIENTS

½ cup peeled mallow "cheeses" (immature, green
seedheads)

1½ cups water

1 egg white (optional but recommended)

¼ teaspoon cream of tartar

¼ cup powdered sugar

½ teaspoon vanilla extract

> First a note about the peeled mallow seed-
> heads. Yes, I know it's time consuming and not
> much fun to peel the tiny rounds. Do it anyway.
> If you leave the husks on, the final product has
> a faint but noticeable rotting vegetable taste
> that you will not enjoy. Measure the mallows
> after peeling them.

INSTRUCTIONS

1. An important factor in mallow foam success
   is timing. You can make the egg white–like
   mallow extract up to 2 days ahead (store in the
   refrigerator). But once you've beaten it into foam
   you need to serve it immediately. Even with the
   optional egg white, which helps stabilize the
   foam, mallow foam will start to separate within
   about 15 minutes. Have your dessert already
   plated when you start beating the egg and mallow extract so that all you have to do is spoon
   some foam on top and serve.

2. Put the peeled mallow "cheeses" and the water
   in a small pot over medium-high heat. Boil until
   the liquid is reduced by at least half. You'll know
   they've cooked long enough when you pour a little
   off a spoon and the consistency is more like raw
   egg white than water. Strain through a fine mesh
   strainer into a small bowl or a container with high
   sides. Set aside the now cooked mallow "cheeses"

for the next recipe. Let the mallow extract cool to room temperature. (It's okay to cheat and put it in the refrigerator to hasten the cooling process.)

3. Use an electric beater to beat the egg white to the stiff peak stage. Beat in the cream of tartar. Start beating in the mallow extract a little at a time. Once the mixture has a snowy white color and fluffy texture similar to meringue, beat in the sugar and the vanilla. Serve immediately as a topping for desserts or fruit.

Mallow Miso Spread

Makes approximately 2 cups

A byproduct of making mallow foam is cooked mallow seedheads. This recipe turns them into a healthy, savory spread. Try it on crackers, in sandwiches, or as a dip for raw veggies.

INGREDIENTS

8 ounces firm silken tofu
½ cup peeled, cooked green mallow seedheads
2 tablespoons miso
2 tablespoons peanut or other light vegetable oil
2 tablespoons rice vinegar
2 garlic cloves, peeled and chopped

INSTRUCTIONS

1. Put all of the ingredients into a blender or food processor and purée until smooth. Note that the mallow will not become completely smooth; the segments of the seedheads will break apart but remain in the otherwise silky purée, providing an interesting texture contrast.

Mint
MENTHA SPECIES

We grow up with mint's scent and flavor in our candies, toothpaste, tea, and dozens of other foods and products. The wild herb is every bit as versatile as its cultivated cousin.

Find
Look for mint on riverbanks, along lake shorelines, and in moist meadows, growing in full to partial sunlight.

Identify
All mints have square stems. This might not be obvious at first glance, but twirl a stem between your thumb and forefinger and you'll immediately feel the four sides.

Mints have opposite leaves (they join the stems in aligned pairs) with toothed margins. They may or may not be aromatic, but only bother with the ones that are (the same essential oils and menthol that give mint its scent give it its flavor. If it doesn't smell like mint, it won't taste like mint).

Depending on the species, the pink, white, or lavender flowers grow in rounded clusters at the leaf axils (where the leaves join the stems) or in elongated clusters at the stem tips.

Sustainably Harvest
Although mint is a notoriously invasive plant in gardens, in the wild it usually is not. If you pinch off the

top few inches of the plants, they will grow back bushier than they were before you harvested from them.

Note that not all mints are created equal when it comes to aroma and taste. Because different mints cross with each other easily, it is not uncommon to come across "new" breeds of mint, some of which are quite flavorless. Use your nose to determine whether you've found a mint worth harvesting.

Eat and Drink

The most obvious use for mint is as an herbal tea. For this use, it is wonderful both fresh and dried, served hot or iced. It is also a fabulous flavoring for cocktails.

You can flavor homemade jelly with mint to create a condiment traditionally served with lamb or game meats.

Although in the West mint is most often associated with candy, gum, and sweet things, in the Middle East, North Africa, Eastern Europe, and Indonesia it is often included in savory dishes.

Mint dries well and can also be preserved by infusing simple sugar syrup with it.

Tabbouleh

Serves 2; recipe can be multiplied, but see note below

Tabbouleh is a classic example of a savory recipe using mint. It is too often made as a grain salad with herbs; the real version is a mint and parsley salad with some grain and vegetables added.

INGREDIENTS

¼ cup medium-fine bulgur wheat

3 tablespoons extra-virgin olive oil

2 tablespoons lemon juice

1½ teaspoons Moroccan-style preserved lemons or 1 teaspoon lemon zest

2 medium tomatoes, cored, seeded, and chopped (about 1 cup once chopped)

1 cup parsley, minced

½ cup fresh mint, minced

½ cup scallions, finely chopped

Salt and freshly ground pepper

INSTRUCTIONS

1. Put the bulgur into any heatproof vessel and pour boiling hot water over it, enough to cover the grain plus about ½ inch of hot water. Let sit 20 minutes.

2. Drain the bulgur in a fine-meshed sieve. Squeeze out excess water (squeeze harder than you'd think).

3. In a large bowl, combine the bulgur with the olive oil and lemon juice. Stir in the other ingredients.

Note: Tabbouleh is best served fresh. The texture of the tomatoes and the herbs suffers when the dish is refrigerated, so only make what you will eat right away.

Mint Julep

Makes 1 mint julep (recipe can be multiplied)

The key to making this quintessential drink of the American South is good-quality bourbon and shaved ice. The shaved ice, stirred, frosts the glass and slightly dilutes the bourbon (but not too much) in a way that ice cubes won't. You can throw some ice cubes into a blender to make the shaved ice if need be.

INGREDIENTS

1 tablespoon water

2 teaspoons sugar

8 mint leaves, plus a few sprigs for garnish

2 ounces bourbon

INSTRUCTIONS

1. In a julep cup or a highball glass, stir together the water and sugar until the sugar dissolves. Add the mint leaves and break them up with a muddler or the handle of a wooden spoon.

2. Fill the glass with shaved ice. Add the bourbon. Stir vigorously to combine the ingredients. When the glass is frosty, garnish with a sprig of mint, drop in a straw, and serve at once.

Morel Mushroom
MORCHELLA SPECIES

Find
Morels are a spring mushroom, and that's part of their ID: Tell me you found a morel in late fall and I'm going to wonder what you really found. Look for them on the ground in fruit tree orchards and at the site of a recent fire.

Identify
Look for morels in April, May, and early June

Where you find one, you are likely to find more, but they aren't always easy to spot. Last year's leaf litter usually partially covers morels and blends in with the gray, black, or brown color of their caps. The caps are most often a darker color than the stems and are covered with a honeycomb-like pattern of pits and ridges. Size can range from less than an inch high to several inches.

There is a similar-looking, inedible mushroom called the false morel (*Verpa bohemica*). It may cause gastric distress if eaten. However, it is very easy to tell morels and false morels apart. False morel caps have rounded, convoluted ridges that look almost like brains, whereas the edible *Morchella* species have a pitted, honeycomb pattern of ridges. Still not sure? Cut the mushroom in half lengthwise: true, edible morels are hollow all the way through.

Sustainably Harvest
You are not harming the mycelium—the buried network that makes up most of the mushroom organism—by harvesting the fruiting bodies (the part we call "the mushroom").

Cut off each morel at approximately ground level with a sharp knife. Soak in a tub of water to remove any insect inhabitants before cooking or dehydrating.

Eat
Simply slivered and sautéed with a little butter or olive oil, morels are wonderful in omelets, risotto, cooked alongside wild springtime alliums such as

ramps or end-of-season field garlic, or on pasta. Keep the preparation simple, because morel flavor is mild and the not-quite-crunchy texture is part of what makes them special. You don't want to bury them in complicated sauces or recipes with too many competing ingredients.

Morels are one of the best mushroom candidates for dehydrating. The rehydrated mushrooms are almost identical to the fresh ones. To rehydrate dried morels, pour boiling water over them and let them soak for 20 minutes. Drain, reserving the soaking water to use in soups or sauces. Treat rehydrated morels like fresh morels in any recipe.

Cleaning morels can be a hassle because of all the tiny crevices. Soaking them in salty water overnight dislodges dirt and any bugs.

Stuffed Morels

Serves 2 to 4

The combination of textures and tastes in these stuffed morels makes them a real treat.

INGREDIENTS

1 pint morel mushrooms

¼ cup (4 tablespoons) butter, bacon grease, or coconut oil

½ cup scallions, chopped

1 large garlic clove, peeled and minced

¼ cup fresh parsley, minced

¼ cup pine nuts or walnuts, lightly chopped or crushed

Salt and freshly ground black pepper

2 tablespoons dry breadcrumbs

¼ cup Parmesan or Romano cheese, grated

INSTRUCTIONS

1. Preheat oven to 400°F.

2. Cut the stems off the morels and set aside. Bring a pot of water to a boil and boil the caps for 10 minutes. Drain in a colander.

3. Chop the morel stems. Sauté in a skillet with the butter, bacon grease, or coconut oil and the scallions for 8 to 10 minutes, stirring often. Add the garlic and cook for 1 minute more, stirring constantly. The liquid should be almost completely evaporated at this point. Remove from the heat and stir in the parsley, pine nuts, and salt and pepper to taste.

4. Poke the stuffing into the morel caps (a chopstick can come in handy here). Lay the stuffed morels in a single layer on a lightly greased baking sheet or on a baking sheet lined with parchment paper. Sprinkle the breadcrumbs and grated cheese over the mushroom caps. Bake for 15 to 20 minutes. Serve hot.

Mugwort
ARTEMISIA VULGARIS

Although destructively invasive in many regions, this is a beautiful and extremely useful herb. It makes a subtle seasoning or a warming aromatic tea, can replace hops in beer making, and has multiple medicinal uses.

Find

Mugwort loves disturbed soils and sunny locations. That means it is frequently found along roadsides, as a farm and garden weed, and in city parks. This is an extremely invasive plant, and it often takes over abandoned lots and neglected yards, turning them into mugwort monocultures.

Identify

Mugwort has deeply divided leaves that are green on the upper surface and white on the underside. The tips of young mugwort leaves may be slightly rounded, but they become sharply pointed as the plant matures. The leaves at the top of the fully grown, 4- to 5-foot-tall stalks are a completely different shape from the leaves lower down on the same stalks. The upper leaves are small, undivided, and linear.

The entire plant has a very distinctive spicy-herby scent when crushed.

Mugwort's tiny gray-green flowers grow in clusters at the top of the stalks. The perennial plants die back for winter, and new plants emerge from the roots. Last year's brown stalks are a good way to locate the plants when they are just beginning to regrow in early spring.

Sustainably Harvest

Seriously? This is one of the most invasive weeds on the planet. It spreads by windblown seed, horizontally spreading rhizomes, and its roots are allelopathic. This means that they exude a chemical that over time actually discourages other plants from

growing near it. Do your neighborhood ecosystem a favor and harvest as much mugwort as you can use.

Eat

Think of mugwort as an herb, not a vegetable. It is fabulous for seasoning food, but the flavor is too strong to eat in quantity. There are commercial brands of mugwort-flavored soba noodles, and that gives a clue as to the Asian-inspired flavors that pair well with it. Mugwort and ginger, for example, is a great combination.

Mugwort can be used fresh or dried. As a seasoning, I think its flavor is best when harvested in midspring, but it is fine to use it later in the growing season as well.

If you decide to sip a mug of mugwort tea, keep in mind its medicinal properties: It will likely make you sweat a little, is a fairly strong muscle relaxer, will bring on menstruation delayed due to stress, and has a reputation for enhancing dream recall (sorry, I haven't noticed that effect from eating the following recipes). Mugwort is also the herb burned as *moxa* in acupuncture

Caution: Women who are pregnant or trying to become pregnant should not eat mugwort.

Artemisia Vinegar

Makes 1 cup

This vinegar seems to act as an MSG-like flavor enhancer—without any of MSG's nasty side effects. Used in a marinade, it ramps up the umami taste of meat. Used in a salad dressing, it makes the individual flavor of each vegetable and other ingredient in the salad stand out.

INGREDIENTS

⅓ cup fresh mugwort leaves, chopped

1 cup rice vinegar

INSTRUCTIONS

1. Rinse the mugwort leaves and put them into a clean glass jar. Pour the vinegar over the leaves. Cover and let steep for 2 weeks.

2. Strain the vinegar into a clean glass bottle. Cover and store for up to 6 months at room temperature or a year in the refrigerator.

Triple Mugwort Soba Noodles

Serves 4

This recipe was inspired by my love of cold sesame noodles plus my discovery of a brand of mugwort-flavored soba noodles. I added fresh mugwort as well as mugwort vinegar to the sauce, but the result is still subtly rather than overly flavored. Regular buckwheat soba noodles will work if you can't find the mugwort-flavored version.

INGREDIENTS

½ cup scallions, chopped

1 medium cucumber

½ cup tahini

2 tablespoons light miso

2 tablespoons soy sauce

2 tablespoons dark (toasted) sesame oil

1 tablespoon Artemisia Vinegar (page 152) or plain rice vinegar

1 tablespoon fresh mugwort leaves, minced

1 teaspoon fresh ginger, grated

½ cup water

8 ounces mugwort-flavored soba noodles (I recommend Eden Foods brand) or plain buckwheat soba noodles

1 teaspoon red pepper flakes (optional)

INSTRUCTIONS

1. Bring a large pot of water to a boil. Also fill a large bowl with cold water. While you're waiting for the pot of water to come to a boil, prepare the vegetables and sauce.

2. Chop the scallions and set them aside. Peel the cucumber and cut it in half lengthwise. Scoop out the seeds with a spoon. Slice crosswise into thin crescents and set aside.

3. Whisk together the tahini, miso, soy sauce, sesame oil, mugwort vinegar, mugwort leaves, and ginger. Alternatively, purée them with an immersion blender or in a regular blender or food processor. Add the water a little at a time until the sauce is approximately the consistency of heavy cream.

4. Put the noodles into the boiling water. Reduce the heat so that the noodles cook at a steady simmer for 5 to 8 minutes (check the package for the specific cooking time for the brand of soba you have).

continued

5. Drain the noodles in a colander, then immediately dump them into the bowl of cold water. Get in there with your clean hands and give the soba a gentle rub. This step is crucial even though the package instructions almost never mention it. It is the difference between ending up with a gluey mass or individual noodles that your chopsticks can pick up easily.

6. Toss the soba noodles with the sauce. Top with the scallions and cucumber. Sprinkle the pepper flakes on top and serve.

Mulberry
MORUS RUBRA, M. NIGRA, AND M. ALBA

I don't know why mulberries are so underappreci-
ated. Although homeowners often curse the falling
berries for staining their sidewalks, and although
some species can become invasive, this is a delectable and versatile fruit.

Find

The red mulberry, *Morus rubra,* is fairly shade tolerant, but *M. alba* and *M. nigra* prefer full sun. In the
wild, look for mulberries in floodplain woodlands.
They are also a common urban and suburban tree.

Identify

Mulberry trees can grow up to 60 feet tall, but they
are usually much shorter than that. The trees have
a scruffy appearance, with the branches sticking
out at odd angles. Seen from a distance, I often
think mulberry trees look like they are having "a
bad hair day."

Often there are three leaf shapes growing on
the same tree: a two-lobed mitten shape, a three-
lobed leaf, and a roughly heart-shaped leaf. Note
that there is another tree out there with those three
leaf shapes: sassafras. But the leaf margins of sassafras are smooth, whereas those of mulberry are
toothed. When there is only one leaf shape on a mulberry, it will be the simple heart shape. No matter
what shape, mulberry leaves grow in an alternate
arrangement.

The bark of mulberry trees develops craggy ver-

tical furrows as the trees age. The branches emerge
from the short trunks just a few feet above the
ground.

As for the fruit, mulberries look very much like
blackberries, although depending on the species, the
fruit may be ripe when it is dark purple or when it is
pale pink. (FYI, blackberries do not grow on trees.
Whenever someone tells me they found a "blackberry tree," I know that what they really found was
mulberries).

155

Mulberry fruits ripen over several weeks, and it is common to see green, white, red, and dark purple (almost black) fruits on one tree at the same time.

Sustainably Harvest

Peak mulberry picking season stretches from late spring through early summer. It's easy to tell when the berries are ripening because they start dropping to the ground.

The quickest way to collect them is to lay down a drop cloth and then shake the branches: The ripe mulberries will fall immediately. If you prefer to pick them off the tree, take only those that yield to your gentle pull without resistance. If you have to tug, that one isn't ripe yet.

Harvesting the berries does not hurt the tree. In fact, you're actually curbing the spread of these sometimes invasive species by gathering their fruits.

Eat

Whether fallen fruit or plucked from the branch, mulberries always come off the tree with a small bit of stem attached. These are a hassle to remove, and often I don't bother. But if you're serving guests you might want to take the time.

The easiest way to do this is to freeze them first: Spread the berries in a single layer on baking sheets and freeze them, uncovered, for an hour or two. Don't just dump fresh berries into a container and freeze them or you'll end up with a berry brick—the single layer freeze prevents that. Pulling off the little stems while the berries are frozen solid is much easier than doing so while they are fresh and squishy.

Mulberries are mildly sweet and pleasant raw. They are also good in pies, ice cream, jam, and homemade wine. If you decide to make mulberry jam, keep in mind that they require some added pectin and acidity.

As with other berries, mulberries freeze well.

Dried mulberries are a worthy ingredient unto themselves. When dried, the mulberry flavor intensifies in a wonderful way. Dried mulberries are great to snack on as is, but they are also wonderful when rehydrated and used in recipes such as this chutney. An added plus is that the little stems break off easily once the berries are dehydrated.

Mulberry Chutney

Makes 1 pint

Sweet and sour with a touch of heat from the red pepper, this chutney is good with any kind of meat, poultry, cooked whole grain, or cheese. Using rehydrated dried mulberries makes for a more intensely flavored condiment than if you use fresh berries, but the chutney is good either way.

INGREDIENTS

2 cups fresh or rehydrated dried mulberries, or a combination of the two

½ cup apple cider vinegar

½ cup apple, finely chopped

⅓ cup honey

¼ cup onion, finely chopped

¼ cup raisins

1-inch piece fresh ginger, peeled and grated

1 teaspoon red chili pepper flakes

1 teaspoon kosher or sea salt

¾ teaspoon ground spicebush berries or ½ teaspoon ground allspice plus ¼ teaspoon ground pepper

¼ teaspoon ground cardamom

INSTRUCTIONS

1. Combine all the ingredients in a pot over medium heat. Simmer, stirring occasionally, until most of the liquid has evaporated or been absorbed.

2. Refrigerate and use within 2 weeks; freeze for up to 1 year; or pack into canning jars, leaving ½ inch headspace, and process in a boiling water bath for 10 minutes. (Adjust canning time if you live at a high altitude. See the Resources section if you're unfamiliar with canning.)

Mulberry Ice Box Cake with Rosewater Cream

Serves 6

This no-bake recipe looks fancy but is super simple to make. The hint of rosewater pairs nicely with the mulberries without overwhelming their gentle flavor.

INGREDIENTS

2 cups heavy whipping cream

2 tablespoons powdered sugar

¼ teaspoon rosewater

3½ cups fresh mulberries, divided

16 to 20 graham crackers, petit beurre cookies, or thin sugar cookies

INSTRUCTIONS

1. Pour the cream into a large bowl. Beat it with a hand mixer or stand mixer until it forms soft peaks. Add the sugar and rosewater and continue to whip the cream until it again forms stiff peaks.

2. Using small ramekins, a muffin tin, or decorative molds, spread about an inch of the cream in the bottom of 6 of the forms. Place in the freezer.

3. Spread a thin layer of the cream on a platter or large plate. Add a layer of the graham crackers or cookies to form a 6-inch square. Spread a layer of the cream on top, then arrange 1 cup of the mulberries on the cream. Cover the mulberries with another layer of crackers or cookies, another layer of cream, and another cup of berries. Repeat one more time, then finish the top with a layer of cookies or crackers and one last layer of cream. You should have four layers of cookies plus cream and three layers of mulberries.

4. Refrigerate the mulberry cake for 2 to 4 hours. To serve, put a slice of the ice box cake on a plate. Top with a few of the remaining ½ cup of mulberries. Put one of the frozen cream circles alongside the cake.

Mustard

BRASSICA NIGRA, B. RAPA, SINAPIS JUNCEA, AND SINAPIS ALBA

Mustard leaves are one of the best cooked greens on the planet, but that is just one of the foods these plants provide. The unripe pods and the flowers are also edible, and of course the seeds make the famous condiment that is used worldwide.

Find

Mustards prefer full sun but sometimes grow where they get only partial sunlight. They are common in unweeded farm fields, gardens, parks, fencerows, and roadsides.

Identify

Mustard leaves start out as a rosette of irregularly lobed, toothed leaves that can grow as large as 20 inches long, although they are usually significantly smaller than that. The leaves on the upper flower stalks are most often not lobed, or at least less so than the rosette leaves. *Brassica rapa*'s upper leaves clasp the stems. Once in flower, mustard plants can grow to be anywhere from 2 to 6 feet tall, depending on the species.

Mustard flowers have four yellow petals in the cross pattern that is one of the hallmarks of the mustard plant family. All mustards have four petals, no exceptions. The flowers grow in clusters, and before they open the clusters of green buds look like miniature versions of the broccoli to which they are related. Look closely and you'll notice that two of

the stamens are shorter than the other four, another characteristic typical of mustards.

Mustards' slender seedpods have a narrow pointed tip sometimes referred to as a "beak."

Every part of the mustard plant has a spicy smell that is something like a cross between a turnip and horseradish. Remember that plant identification isn't all about the visual: use your nose!

Depending on when the plants germinate, mustards may behave like annuals—flowering, going to seed, and dying within a year—or they may act like biennials and overwinter before going to seed the following year.

Sustainably Harvest

Mustard plants can be invasive, so no worries about harvesting them. However, if you find a particularly good patch, be sure to leave a few plants to go to seed and generate next year's wild crop.

Gather mustard leaves starting in early spring. Mustard greens do get more bitter as the season progresses, but some species are milder than others and can be harvested even in early summer.

Collect the unopened bud clusters and the flowers from late spring through early fall, depending on the region. Nibble on the seedpods while they are still tender and green.

The seeds are ready to harvest once the seedpods have dried on the plants in summer and fall. The best way to harvest them is to put a bag over them and break off the stalks (with the seedheads attached) directly into the bag. Let them dry for a few days, then crush the seedpods by rolling the bag with a rolling pin or a wine bottle. Transfer the mustard seeds and chaff to a large bowl in front of a fan or outside on a windy day. Lift a handful of seeds and chaff: the seeds will drop back into the bowl while the lighter chaff blows away.

Eat

Young mustard greens are good raw in salads, especially when mixed with blander leafy greens.

Cooked, they are pleasantly spicy and good on their own with a little garlic and olive oil, or mixed with milder greens such as quickweed or Asiatic dayflower. They are also good in strongly seasoned dishes such as chili and curries.

Mustard greens can be blanched and frozen for future use.

Add the broccoli-like immature flower clusters to other cooked wild vegetables. They are tasty enough to eat on their own, but too small to make it worth harvesting them in large quantities. Some foragers have found them disappointingly bitter, but I have not had that experience. The immature flower clusters can be boiled, steamed, or stir-fried.

Once the spicy flowers open, use them as an attractive and tasty salad garnish. Or add them to Vietnamese-style transparent spring rolls or any other dish that shows off their color as well as their flavor.

Use mustard seeds whole in curries and in pickling spice blends. Or grind them and add water, vinegar, beer, or another liquid to make your own prepared, spreadable mustard. Be sure to let your mustard age for a few weeks, because freshly made prepared mustard can have a bitter taste that mellows after a couple of weeks. Homemade prepared mustard will keep in the fridge for 1 year. The whole seeds can be kept at room temperature for at least a year if they are kept in a tightly covered container.

Pickled Mustard Leaves

Makes 1 pint

Pickled mustard leaves are popular in China and Taiwan, and
you may have eaten them without
realizing it. They are often served with
soup, pork rice, and many other dishes.

Recipes for making pickled mustard leaves
vary from kimchi-style ferments to a strongly
sour vinegar pickle. This recipe is for a sweet
and sour version that is good served as a condiment alongside Asian-style meals. It will also
jazz up a simple veggie soup.

The mustard leaves may lose some of their
bright green color and darken a bit once they've
sat in the brine for a few days. This is normal
and does not affect their flavor.

*Note: This recipe is not good for canning. The brine is not
acidic enough to safely process in a boiling water bath. You
could pressure can it, but the greens would turn to mush.
Stick with refrigeration for this one.*

INGREDIENTS

1 pound fresh wild mustard leaves (feel free to include
some of the immature green flower heads, too)

2 cups water

½ cup sugar or ⅓ cup light honey

2 teaspoons non-iodized salt

½ teaspoon whole black peppercorns

½ cup rice vinegar

INSTRUCTIONS

1. Wash and coarsely chop the mustard greens (you
can leave the pieces fairly large). Loosely pack the
greens into a clean, glass pint jar.

2. Bring the water, sugar (or honey), salt, and
peppercorns to a boil, stirring to dissolve the
sugar and salt. Remove from heat and stir in the
vinegar.

3. Pour the brine over the mustard greens. Press
down on the greens with the back of a spoon to
release any air bubbles and settle the leaves down
into the brine. Cover and store in the refrigerator.
Wait at least a week before sampling.

Spicy Sprouted Mustard and Avocado Salad

Serves 2

Sprouted mustard seeds (instructions after the recipe) and a mustard-based dressing make that plant the star of this salad. Creamy avocado balances it. Add the chili peppers if, like me, you like it hot. Leave them out for a milder variation.

INGREDIENTS

1 pint sprouted mustard seeds

½ medium-size, ripe avocado, sliced lengthwise

1 medium tomato, sliced

1 small chili pepper, stem end and seeds removed, thinly sliced (optional)

4 tablespoons extra-virgin olive oil

2 tablespoons freshly squeezed lemon juice

1 tablespoon Dijon or homemade prepared mustard

½ teaspoon salt

¼ teaspoon dried thyme

¼ teaspoon freshly ground black pepper

⅛ teaspoon liquid smoke or ground chipotle peppers (optional)

INSTRUCTIONS

1. Divide the sprouts between two bowls and arrange them so that they form little nests. Arrange half of the avocado and tomato slices in the center of each mustard sprout nest. Arrange the chili pepper slices, if using, over the top.

continued

2. Whisk together the oil, lemon juice, mustard, salt, thyme, pepper, and liquid smoke or chipotle (if using). Drizzle over the vegetables and serve immediately.

HOW TO SPROUT WILD MUSTARD SEEDS

1. Put 3 tablespoons of wild mustard seeds in the bottom of a wide-mouth pint jar. (Believe it or not, those 3 tablespoons will morph into almost a full pint of sprouts, so don't be tempted to add more.)

2. Fill the jar with water. Cover the top with a piece of cheesecloth, butter muslin, or screen and secure with a rubber band or string. Let the seeds soak 8 to 12 hours or overnight.

3. Drain the seeds in a fine mesh sieve and rinse them under cool water. Put them back into the jar still wet from their rinse, but don't add any additional water at this time. Cover with a breathable material as before, and place somewhere away from direct light or heat.

4. Twice a day, rinse the seeds and return them to their jar. Around day 2 or 3, you'll see the seeds start to germinate. They look a bit like quinoa at that stage. Soon they'll start looking more like alfalfa sprouts. When they reach the alfalfa sprout stage, move them into a sunnier spot for a day. Keep an eye on them to make sure they don't dry out.

5. On day 4 or 5, dump the sprouts into a bowl and cover them with cool water. Untangle them somewhat with your fingers and swish them around to release the seed hulls. These will float to the top, where you can skim them off with a spoon. Repeat with a few changes of water. You won't get 100 percent of the hulls, but by removing most of them you will greatly improve the mouthfeel of your final product.

6. Drain the mustard sprouts well and transfer them to a covered container. Store in the refrigerator and use within 5 days.

Nettles

URTICA DIOICA AND OTHER *URTICA* SPECIES

Nettles are one of the very best wild edibles, both for flavor and nutritional value. And don't worry: the sting factor vanishes when they are dried or cooked.

Find

Nettles love moisture and nutrient-rich soil. They are happy in a variety of light conditions from full sun to partial shade, but won't grow in full shade. Look for them near rivers and streams, in roadside ditches, and in farmland hedgerows.

Identify

Nettles have leaves with toothed edges and pointed tips, and somewhat square stems. These characteristics remind many people of mint, but nettles don't have a noticeable aroma. What they *do* have is stinging hairs. Brush the inside of your wrist on one of the plants and you will have instant confirmation of whether or not it's a stinging nettle. (Don't worry, the sting goes away.)

The usually unbranched stalks can grow as tall as 6 feet, but are much shorter than that during their prime, preflowering harvest stage.

Nettles' tiny, gray-green flowers dangle from the leaf axils like stringy, branched earrings. They're inconspicuous but worth learning to recognize, because once nettle flowers its main harvesting season is over.

Sustainably Harvest

If you pinch off just the top several inches of nettles plants, they will regenerate. To do that, though, you'll need to brave the sting factor. The easiest way to do that, of course, is to wear gloves. But what if you come across a lush patch of nettles at the perfect harvesting stage and you don't have any gloves? Truth is, I often pick them barehanded, and yes, I do get stung. But the sting isn't harmful and goes away, and the harvest is worth it. If you like, you can try the old folk remedy of "nettles in, dock out": Rub a *Rumex* species dock leaf on the stinging area after a nettles encounter, or better yet, apply a spit poultice. A quick rinse with cold water also helps.

You can also minimize the sting factor by carefully grasping the stems from the bottom up rather than the top down. This helps because the stinging hairs on nettles' stems grow either pointing upward or straight outward.

Early to midspring is usually a perfect time to collect nettles, after the perennial plants emerge from winter dormancy but before they flower. It's okay to eat just the young growing tips of the plants later in the season, though.

Eat

A rinse in cold water is enough to lessen the sting, but before you bring nettles to the table they should be either thoroughly dried or cooked.

Boiled or steamed, nettles' leaves are a fantastic leafy vegetable. They have a deep, vibrant green color, meaty texture, and are loaded with vitamins, minerals, and even protein.

I like to dry plenty of nettles' leaves to go into winter infusions. Considered one of the most beneficial tonics in herbal medicine, nettle tea is packed full of nutrients. Its flavor is a bit bland, though, so I usually combine it with mint or some other aromatic herb.

Another way to use dried nettles is to add them to soups, stews, and grain and bean dishes. Just crumble them in while your food is cooking. The flavor will be fairly neutral and won't interfere with the recipe, but the nettles will add a nutritional punch.

Nettles Malfatti

Serves 2 to 3

***Malfatti* means "poorly made" in Italian.** Similar to gnocchi in texture, malfatti was considered famine food, perhaps because it incorporated weeds from the field (nettles in this case) and stale bread. Or perhaps the "poorly made" part has to do with its easy-to-make shape that requires no equipment other than your hands. In any case, malfatti is delicious, and more than other wild greens would, nettles give this dish a wonderful deep green color.

INGREDIENTS

4 ounces raw nettle leaves, stripped from their stems (yeah, you might want to wear gloves for this job)

2 teaspoons olive oil

1 small onion, peeled and finely chopped

2 eggs, beaten

¾ cup dry breadcrumbs

½ cup Parmesan cheese, grated

1 teaspoon salt

¼ teaspoon ground black pepper

INSTRUCTIONS

1. Bring a medium-size pot of water to a boil. Add the nettles and blanch for 3 minutes. Pour into a colander, then immediately run the colander under cold water to stop the cooking. Drain again, then squeeze hard to get out as much water as possible. (Don't worry, the sting factor has already been neutralized by the blanching.)

2. Coarsely chop the wad of blanched, chilled, and squeezed-out nettles.

3. Heat the olive oil in the same pot you used to cook your nettles. (I'm just trying to reduce the number of dishes you need to do later.) Over medium heat, cook the onion in the oil, stirring often, for 5 minutes.

4. Put the onion and the nettles into a food processor with a little salt and pepper. Process until very finely chopped. If you don't have a food processor, just keep chopping until you have something close to a paste.

5. Put the onion and nettles mixture into a large bowl along with the eggs, breadcrumbs, cheese, salt, and pepper. Mix until thoroughly combined.

6. Cover and refrigerate for 6 to 8 hours.

7. Lightly flour a baking sheet and your hands. Pinch off about a tablespoon of the chilled mal-fatti dough and roll it into a torpedo-like shape about 1½ inches long and ¾-inch wide. Place it on the floured baking sheet. Repeat with the rest of the dough.

8. To cook, bring a pot of water to a boil. Pile the malfatti onto a slotted spoon and gently lower them into the water (it may take a few spoonfuls). Boil for 4 to 5 minutes, then use the slotted spoon to transfer the malfatti to a plate or large bowl. Gently toss with melted butter or good-quality extra-virgin olive oil, and top with more grated Parmesan. Serve immediately.

Oak/Acorns
QUERCUS SPECIES

Despite taking some time and effort to prepare, acorns have been a staple food of many cultures. They are nutritious, taste wonderful, and—depending on how you process them—yield two very different ingredients.

Find

Oaks often grow as one of the predominant species in mixed deciduous tree forests. Smaller, shrublike species can be found on sunny hillsides, often in rocky soils. Oaks are also commonly planted in parks and as backyard and street trees.

Identify

Oak species range from tall, stately trees with trunks much bigger than anything this tree hugger can wrap her arms around, to much shorter, shrublike plants.

Among the characteristics they all have in common are alternate, leathery leaves that are usually but not always lobed. The leaves of the *Quercus* species commonly referred to as "white oaks" have rounded lobes. Red and black oaks have sharply pointed lobes. There are also *Quercus* species with unlobed but toothed, holly-like leaves.

Oak's yellow-green male and female flowers dangle like lumpy threads. Usually described as "inconspicuous," you'll notice them most when they are piling up on the ground under the tree.

The acorns that we are interested in for food are

nuts that are sometimes shaped like 2-inch-long and ½-inch-wide bullets, sometimes round and an inch or more in diameter, other times small and squat. Whatever shape they take, all acorns have thin, brown shells nested in a detachable "cup" or cap.

Sustainably Harvest

Gather acorns soon after they've fallen to the ground in late summer or early fall. Don't wait too long or the good ones may disappear to your competition—mostly squirrels.

Inspect each acorn for small holes. These are a sign of a weevil's exit, and there won't be anything worth eating inside. Acorns that dropped to the ground while their shells were still green or whitish are also not worth bothering with.

Oak trees don't produce acorns consistently every year. They have what are called "mast years" in which they produce copious quantities of nuts.

The next year they may produce less than half as many. This on-again, off-again pattern is thought to be an evolutionary survival mechanism.

Eat

Acorns are excellent eating . . . once you've gone to the trouble of gathering, shelling, and leaching them. Unprocessed, the high tannin content of most acorns (some oaks produce milder acorns than others) makes for a strong, unpleasant astringency. But once those tannins have been leached out, acorns are transformed into a delicious, nutrient-dense food that can be enjoyed in many different ways.

Two Ways to Process Acorns

First, put your acorns into a big bowl of water. The good acorns will sink to the bottom. Discard any floaters; bugs got to them before you did.

Next, use a nutcracker or a small hammer to shell the acorns. The shells are thin and not too difficult to remove.

Okay, you've weeded out the good acorns from the buggy, bad ones, and you've shelled your haul. Now what?

There are two basic methods for leaching the tannins out of acorns: the cold water method and the hot water method. They are both effective, but they yield two completely different acorn flours. There's also a third product, whole acorn nuts, that uses the hot water method, so let's start with that method.

The Hot Water Method

Put shelled acorns into a large pot and cover them with water. Bring the water to a boil over high heat. The water will darken from the tannins leaching out of the acorns. Drain in a colander, return the acorns to the pot, cover with fresh water, and bring to a boil again. It helps speed things up if you have a second pot of water coming to a boil while you are leaching the acorns in the first pot. Keep boiling the acorns in fresh changes of water until the water no longer turns dark brown. If you aren't sure if the acorns have been leached enough, just nibble on one: it should taste mild, not at all bitter or astringent.

> Tip: You don't have to do the hot leaching process all at once. You can spread it out over several days, boiling the acorns and changing the water when you're home, then turning the stove off and getting back to it later.

Hot water–processed acorns can be eaten whole or ground into meal. A food processor is helpful for the grinding. Spread the ground acorn meal out on baking sheets and dry it in a low oven, or do the same in a dehydrator. (Line the trays with silicon sheets or parchment paper first.) Use dried acorn meal as is in hot cereals and baked goods, or regrind it in a food processor, blender, grain mill, or coffee grinder to get a finer flour.

Hot water–processed acorn flour has a dark,

toasty color and adds a crumbly texture to baked goods. Remember that acorns are gluten-free and won't help homemade bread rise: I usually add no more than 25 percent hot water–processed acorn flour to wheat flour if I'm making bread with hot water–processed acorn flour.

The Cold Water Method

Cold water–processed acorn flour has a lighter color, and although it is also gluten-free, it can create a spongy texture in baked goods that hot water–leached acorn flour cannot. For quick breads made with baking soda or baking powder, you can use 100 percent cold water–processed acorn flour (with no other type of flour added), although it does help to include egg or flax seed meal in the dough as a binder.

To cold-process acorns, it's best to grind the nutmeats before you start leaching them. Again, a food processor comes in handy here. Back in the day, you would've put your ground acorns into a permeable woven bag or basket and attached it to the bank of a cool stream. The water would have run through your acorn meal for a few days, leaching out any tannins. If you try this, be on guard for other acorn lovers such as squirrels, who will be happy to make off with your bag of acorn meal. (I speak from experience.) Fortunately, there are other ways to cold water–leach acorns.

Many foragers use the clean water in the tank at the back of their toilets to leach acorns (yes, you read that correctly). If you have the kind of toilet that has an accessible clean water reservoir, you can use that (the reservoir, *not* the bowl). Disinfect the toilet reservoir by adding ¼ cup liquid bleach, letting it sit for an hour, and then waiting until the toilet has been flushed enough times to rinse the reservoir free of any bleach smell.

Put the ground acorn nutmeats into a cloth bag. Tie the bag tightly shut and put it in the water of the toilet tank. Every time someone flushes, the tank will empty and refill, mimicking that cold water stream I mentioned. Once the water stays clear, take out the bag of cold water–leached acorn meal.

If the toilet-leaching thing is not for you, you can also cold water–leach acorns by stirring the acorn meal together with cold water in a big pot. Leave that in the refrigerator overnight, then drain off the liquid or strain through butter muslin, several layers of cheesecloth, or a jelly bag. Return the acorn meal to the pot with fresh cold water and put it back in the fridge for another 8 to 12 hours. Repeat until a pinch of the acorn meal no longer has any mouth-puckering feel when you taste it. Note that this method can take as long as 2 weeks, depending on how high in tannins the particular batch of acorns is.

Whichever method you use, at the end of it you'll have some wet, leached acorn meal with the texture of coarse polenta. Wrap it up in the cloth you strained it through and squeeze hard to remove as much liquid as possible. Spread the resulting mush out on baking sheets and place in the oven on the lowest possible setting. Better yet, put it on dehy-drator sheets and set the dehydrator to around 95°F. Stir the acorn meal occasionally while it dries (which may take anywhere from an hour to a day, depend-ing on how thickly you spread it and the temperature at which you're drying it). The meal needs to be 100 percent dry for storage.

You can use the meal as is, but for the recipes below you need to grind it into a fine, powdery flour. Your food processor is not up to the job, but an elec-tric coffee grinder works great, as does a Vitamix.

Acorn flour goes rancid if stored for long at room temperature, and after all the work you put into it, you don't want that. Store acorn flour in the refrig-erator or freezer for up to 2 years.

Acorn Blinis

For this recipe it is important to use only cold water–processed acorn flour. It gives the blinis a lovely, spongy texture that hot water–processed acorn flour could not, especially since there is no other flour in this recipe.

It is especially important to grind the flour for this recipe until it is very fine, close to a powder. As mentioned, if your food processor isn't doing the job, an electric coffee grinder will.

INGREDIENTS

1 cup cold water–processed acorn flour
2 tablespoons butter, melted
1 egg, beaten
½ cup milk
¼ cup buttermilk
¼ teaspoon salt

INSTRUCTIONS

1. Mix all of the ingredients together in a bowl. Let the batter sit at room temperature for 10 minutes so that the acorn flour can absorb the moisture from the other ingredients. Add additional milk, if necessary, to make it approximately the consistency of pancake batter.

2. Place a griddle or frying pan over medium-low heat. The pan is hot enough when a drop of water sizzles immediately on contact.

3. Drop no more than a tablespoon of batter on the griddle for each blini. It's fine to have several cooking at once as long as there is enough space between them to easily slide in your spatula.

4. Let the blinis cook on the first side until the edges dry slightly and the bubbles that form have almost all popped. Flip them over and cook them on the second side until they slide around easily when you shake the pan, a sign that they are browned on the second side.

Like crepes, blinis can go in either savory or sweet directions. But if you want to serve them up in traditional style, break out the caviar and some sour cream or crème fraîche.

Acorn Focaccia

Note that this is an overnight recipe, so plan ahead. The long, cold rise in the refrigerator gives the focaccia a much richer flavor and better texture than a quicker, room temperature rise would.

INGREDIENTS

2 teaspoons active dry yeast

1 teaspoon honey

1½ cups lukewarm water

4 tablespoons extra-virgin olive oil, divided, plus additional for greasing a mixing bowl

2¾ cups all-purpose flour or bread flour, plus additional for dusting work surface

¾ cup acorn flour made from either hot water– or cold water–processed acorns (you'll get a very different focaccia depending on which you use, but both are good)

1½ teaspoons salt

INSTRUCTIONS

1. Dissolve the yeast and honey in the water and let it proof for a few minutes. (If the yeast is good, the surface of the liquid should get frothy.) Add 3 tablespoons of the olive oil.

2. If kneading by hand, whisk together the flour, acorn flour, and salt. Add the dry ingredients to the wet ingredients, ½ cup at a time, until you can form a ball of dough that is easy to transfer to a lightly floured work surface. Knead the dough for 10 minutes, adding the flour mixture a little at a time as necessary. Don't add too much of the dry ingredients; the dough should still be sticky.

3. If using a stand mixer, first combine the dry ingredients in a separate bowl. Add half of the dry ingredients to the work bowl of the mixer with the dough hook attachment. Add all of the liquid ingredients. Let the machine knead the dough at medium speed for 8 to 10 minutes, adding more of the dry ingredients a little at a time. The dough will eventually form a ball around the hook. It should still be sticky, though. If in doubt, err on the side of adding too little rather than too

continued

much of the dry ingredients. (You can save any unused dry mixture in the freezer for future acorn focaccia.)

4. Lightly grease a large bowl with olive oil. Put the ball of dough into it, then flip the dough over so both sides are greased. Cover the bowl with a clean, damp dish towel or a plate. Let it rise for 20 minutes at room temperature, then transfer it to the refrigerator for 8 hours or overnight.

5. Take the dough out and punch it down. Knead it for a couple of minutes. Let it rest at room temperature for 15 minutes.

6. Preheat the oven to 425°F.

7. Line a baking sheet with parchment paper. Lightly grease the paper (cooking spray works here). Press out the dough until it is almost the size of the baking sheet. Do this gradually, rolling or pressing out the dough, then letting it rest for 5 minutes so the glutens can relax. Repeat the rolling out/5-minute rest sequence a few times until the dough is stretched to the right size.

8. Cover with a clean, damp dish towel and let rest at room temperature for 30 minutes. Use your fingers to press dimples all over the dough. Baste the surface of the dough with the remaining 1 tablespoon of olive oil. Top with any toppings you like—herbs,

sea salt, sumac powder, sliced olives, and thinly sliced tomatoes are all good options.

9. Bake for 20 to 25 minutes until turning golden. Let rest for at least 10 minutes before serving.

Oyster Mushroom
PLEUROTUS OSTREATUS

The oyster mushrooms you buy in markets are cultivated, but fortunately for foragers this is one of the most abundant and choice edible wild mushrooms. I have even found oysters in the middle of winter!

Find

Pleurotus ostreatus is a parasitic mushroom that causes a white rot on conifers and hardwoods, but is more likely to be found on hardwoods. Especially common on elms, overlapping clusters of oyster mushrooms are often spotted on elm stumps and logs and on dying and dead elm trees.

Identify

Oyster mushrooms get their common name from the shape of their fruiting bodies. They look a little like stacks of oyster shells, except that they have a soft texture. Unlike a number of other shelf fungi, the undersides of oysters have gills rather than pores.

When present, the stems are off center, with the gills running most of the way down them to the stem's point of attachment. The tops of the caps range in color from off-white to beige, brown, or gray.

Oyster mushroom clusters can be hefty, 20 pounds or more.

Random fact: Oyster is one of the few carnivorous mushrooms. Its mycelia are capable of killing and digesting nematodes.

Sustainably Harvest

Use a sharp knife to cut off the oyster mushrooms near the base. You are leaving the mycelium in the wood, and it will often produce several more flushes of mushrooms.

Eat

Thomas Jefferson used to serve oyster mushrooms braised in cream on toast points as part of his Thanksgiving feasts at Monticello, and he was on the right track with that. Their mild flavor is at its best in creamy sauces or simply sautéed in butter or oil.

"Oyster" Stew

Serves 2 as a main course, 4 as a side dish

This is a playful riff on the seafood stew that is traditionally served at Christmas. The taste is luxurious even though the ingredients are simple. You can use either fresh or dehydrated oyster mushrooms in this recipe with equally good results.

INGREDIENTS

3 tablespoons butter

8 ounces fresh or rehydrated oyster mushrooms, chopped (see note)

2 shallots, minced

¼ cup dry sherry

¾ cup milk

½ cup fish stock or clam juice (use oyster mushroom stock for a vegetarian version)

¼ cup cream

Salt and pepper

1 sprig fresh parsley, minced (optional)

INSTRUCTIONS

1. Melt the butter in a medium-size pot. Add the oyster mushrooms and the shallots and cook over low heat, stirring, until the mushrooms first release and then reabsorb their liquid.

2. Add the sherry and increase the heat to medium-high. Cook, stirring, for 1 minute.

3. Add the milk and fish stock or clam juice and bring to a boil. Reduce heat and simmer for 5 minutes. Remove from heat.

4. Stir in the cream and season with salt and pepper to taste. Serve hot, with the parsley sprinkled on top of each serving.

If you are using dried oyster mushrooms, rehydrate them by pouring boiling hot water over them in a bowl and letting them soak for 15 minutes. Save the liquid to use as mushroom stock. Do not use stock made from other types of mushrooms for this recipe because that could overpower the delicate flavor of the oyster mushrooms.

Peppergrass

LEPIDIUM SPECIES

The edible leaves, seedpods, and flowers of this common weed are all mildly spicy. If you like mustard and arugula, you'll like peppergrass.

Find

Look for peppergrass at the edges of lawns or roads and in parks, fields, meadows, and other sunny locations with disturbed soils.

Identify

Peppergrass, also known as poor man's pepper, is a member of the mustard family. Its lower leaves are up to 3 inches long and lobed. The plants can grow to be a foot high. The leaves on the upper parts of the plant are linear and only an inch long or less. They usually have teeth along the margins or just at the tips. Fine, short hairs grow on the branching stems.

The seedheads grow at the tips of the stems, frequently with a few of the tiny, four-petaled white flowers at their tips. The seedpods are small, flat discs that are notched on one side. The overall appearance of the seedheads resembles a bottlebrush.

Peppergrass grows from branched taproots.

Sustainably Harvest

Peppergrass is considered an invasive weed, and you are not doing any damage by harvesting it. With that said, if you want to encourage it in a particular location you are field managing, simply allow one plant's seeds to mature.

All of the aerial parts of peppergrass are edible, but the stems are usually too tough to use. Gather the leaves anytime. Pick the seedheads when the flat, round pods are still green—once they turn tan they've lost most of their flavor. Usually some of the flowers will come along with your seedpod harvest, and that's fine because they are tasty, too.

Eat

Enjoy peppergrass leaves raw or cooked. Pinch off the tips of the seedheads where a few of the tiny white flowers are attached and use as a salad garnish.

Use the green seedpod disks fresh or dried as a spice. Strip the seedpods off the stems before using. The dried ones only keep their full flavor in storage for 2 to 3 months. For longer storage, freezing the fresh seedpods is a better option.

Peppergrass Chermoula

Makes 1 cup

Chermoula is a spicy marinade that is used to season seafood in Morocco and Tunisia. Peppergrass's mustardy bite fits in perfectly. Don't limit this chermoula's use to fish, as it's also good with grilled vegetables or chicken.

INGREDIENTS

2 cups lightly packed coriander leaves (cilantro)

1 cup lightly packed parsley leaves

¼ cup peppergrass leaves

2 tablespoons fresh green peppergrass seedpods, stems removed

4 garlic cloves, peeled and roughly chopped

1½ teaspoons ground cumin

1 teaspoon sweet paprika

¾ teaspoon salt

¼ teaspoon ground coriander seeds

⅛ to ¼ teaspoon cayenne pepper

½ cup good-quality extra-virgin olive oil

3 tablespoons lemon juice

1 teaspoon minced preserved lemon (optional)

INSTRUCTIONS

1. Put all of the ingredients into a blender or food processor and pulse until you have a coarse purée. You will need to scrape down the sides of the blender or food processor with a spatula a few times to make sure all of the ingredients are incorporated.

2. Use your chermoula as a marinade for at least an hour before cooking fish or chicken, or serve it as a sauce alongside grilled vegetables. Chermoula is at its best freshly made, but you can pour a layer of olive oil over the top of it and store it in the refrigerator for up to 1 week.

> Alternatively, you can make this chermoula by hand. Use a mortar and pestle to mash the garlic with the salt. Mince the cilantro, parsley, and peppergrass leaves. Combine all of the ingredients.

Pineappleweed
MATRICARIA DISCOIDEA AND
M. MATRICARIOIDES

Pineappleweed's flowers and leaves smell and taste like a fruitier, richer version of chamomile. They are delightful in both cold and hot teas, cordials, and syrups.

Find
Look for pineappleweed in sunny, open places such as lawns and roadsides. I also find it frequently at the bases of city street trees, but that, alas, is not a location I want to harvest from due to dog zone and traffic pollution.

Identify
Pineappleweed looks a lot like its close relative, chamomile. Its flower heads are ¼ to ½-inch yellow-green orbs or rounded cones. Each of those small flower heads is made up of many tiny flowers that you can see if you rub one of the flower heads: It will break apart into the individual flowers. Pineappleweed does not have the white, petal-like ray flowers around the yellow-green center that chamomile flowers have.

Although pineappleweed can grow up to 1½ feet tall, it's not unusual to find it growing much smaller, even as short as 2 inches high. This is usually because pineappleweed loves to grow in lawns, where it gets mowed.

Pineappleweed leaves are very finely dissected, which gives them a feathery appearance and feel.

The whole plant is beautifully aromatic when crushed. The scent is unmistakable once you get to know it, like a cross between chamomile and, yeah, pineapple. Using your nose is an important part of confirming your identification of pineappleweed.

Sustainably Harvest
Pineappleweed is, as its common name suggests, considered a weed and is not in any way endan-

gered. However, if you find a particularly lush patch of it that you want to encourage, prune off just the top few inches of the flowering plants. They will rejuvenate lusher than they were before.

Eat

Fresh or dried, hot or iced, pineappleweed leaves and flowers (especially the flowers) make gorgeous teas.

They are also lovely in syrups that can then be used in drinks (alcoholic or otherwise), on fruit salads and desserts, or to sweeten a glass of pineappleweed tea for a double dose of aromatic goodness.

Pineappleweed Cordial

Makes 1½ cups; recipe can be multiplied

To quote my dad, "This is the best cocktail I have had in a long time."

INGREDIENTS

¾ cup fresh pineappleweed flowers and leaves, divided

1½ cups vodka

3 tablespoons light honey such as clover blossom or any light, not too strongly flavored honey (you could also use agave nectar)

INSTRUCTIONS

1. Put ½ cup plus 2 tablespoons of the fresh pineappleweed into a clean glass jar. Pour the vodka over the pineappleweed and cover the jar.

2. Put the remaining 2 tablespoons of the pineappleweed into a small glass jar. Pour the honey over the pineappleweed and cover the jar.

3. Place both jars in a warm, sunny spot for 8 to 24 hours. A sunny window will do if you don't want to leave the jars outdoors.

4. Strain the honey through a fine-mesh sieve into a jar or, preferably, something with a pour spout. The warmth of the sun will have liquified the honey, making it easier to strain. But don't worry if some honey is still gumming up the sieve. The next step will take care of that.

5. Strain the infused vodka through the same sieve you used to strain the honey. The vodka will dissolve any remaining honey on its way through.

6. Transfer to a bottle and cork or seal tightly. Serve chilled.

Pineappleweed Flan

Serves 6

Both the custard base and the sweet coating on this silken dessert are infused with pineappleweed's fruity aroma and taste.

INGREDIENTS

1½ cups pineappleweed sprigs, divided

1½ cups sugar, divided

2 tablespoons water

2 tablespoons lemon juice

2 cups half-and-half

1 teaspoon vanilla extract

3 eggs

2 egg yolks

Pinch salt

INSTRUCTIONS

1. Tie ½ cup of the pineappleweed sprigs at one end with kitchen string or unwaxed dental floss. Put 1 cup of the sugar, the pineappleweed bundle, and

continued

the water in a heavy-bottomed pot over medium-high heat. Cook, stirring occasionally, until the sugar melts. Continue to cook, tipping the pot to mix the syrup (don't use a spoon) until the syrup just begins to darken, about 10 minutes.

2. Remove from the heat and take out the pineappleweed bundle. Immediately add the lemon juice and swirl the pan again to mix. Divide the mixture between six ramekins. Tilt each ramekin so that the syrup evenly coats the bottom and a bit of the sides. Place the ramekins with the pineappleweed syrup in a baking dish.

3. Preheat the oven to 325°F and bring a kettle or pot of water to a boil.

4. Whisk together the half-and-half and vanilla in a small pot over medium-low heat. Tie the remaining cup of pineappleweed sprigs into two bundles and add to the other ingredients. Bring to a simmer; do not let the mixture boil. Remove from the heat and let sit for 5 minutes. Remove the pineappleweed bundles.

5. Place eggs and egg yolks into a large bowl with the remaining ½ cup of sugar and the pinch of salt. Whisk until the mixture is thick and pale. Gradually whisk in the hot half-and-half mixture, starting with just a tablespoonful at a time. (Don't add too much of the hot liquid at once or you'll end up with scrambled eggs.) Once you've added about half of the hot dairy mix, you can stop being so gradual and pour the rest in steadily, whisking the whole time.

6. Strain the mixture through a fine-meshed sieve into a large measuring cup or something else that is easy to pour from. Pour the custard into the syrup-coated ramekins.

7. Pour the hot water into the baking dish. It should come halfway up the side of the ramekins. Carefully move the baking dish into the oven. Bake for 20 to 30 minutes. The custard should still jiggle a bit if you move the ramekins. Let the flans cool in the water bath, then refrigerate for at least 4 hours (or as long as overnight) before unmolding.

8. To unmold, run a knife around the inside of the ramekin to dislodge the flan. Put a plate on top of the ramekin and invert it to release the flan.

Plantain

PLANTAGO MAJOR, P. LANCEOLATA, AND P. RUGELII

No relation to the tropical, banana-like plant with the same common name, plantain is a common weed with edible leaves and seeds. It is also one of the best herbal remedies for scrapes, cuts, bug bites, and bee stings.

Find

If you have a sunny driveway, you probably have some plantain growing at its edges. *Plantago* species love sunny places with disturbed soils and are common in lawns, parks, and other outdoor human habitats.

Identify

All plantains have leaves that grow in a low rosette, and those leaves have prominent, stretchy, parallel veins. If you pull one of the leaves off the plant, you'll often see those veins sticking out of the stalk like threads (think celery). The leaves have smooth edges or a few soft teeth.

Plantago major (common plantain) has wide, oval leaves. *P. rugelii* (Rugel's plantain) leaves are the same shape as common plantain's, but with red or purplish coloration at the base of the leaf stalks. *P. lanceolata* (narrow-leaved or English plantain) has narrow leaves that can grow anywhere from a few inches to a foot long, but are almost never more than an inch wide.

All three species have flowers and seedheads that emerge from the center of the leaf rosette on leafless stalks. *Plantago lanceolata* has 1- to 2-inch long seedheads with tiny white flowers. The seedheads of both *P. major* and *P. rugelii* are several inches long and cover most of their stalks. They start out with green, scale-like seeds which eventually turn brown (*P. major*) or black (*P. rugelii*).

Sustainably Harvest

These are invasive plants and sustainability is not an issue with them. Harvest the young leaves spring through fall. Harvest the seeds after they've turned brown or black. Snip or pinch off the stalks and dry the seedheads for a few days in cloth or paper bags. Strip the seeds off the stalks by holding the tip of each one and stripping back toward the base. I don't bother trying to winnow the chaff from the seeds—just think of it as extra fiber.

Eat (and Use for First Aid)

Use the smaller, tender leaves from the center of the rosettes raw in salads. Use the larger outer leaves to make chips (they're too stringy for other culinary uses). Add the seeds to crackers, breads, muffins, etc.

Although this is a book about edible plants and not herbal medicine, I'd be remiss if I didn't mention plantain's first aid uses because it is so useful when you're in the field. The leaves are anti-inflammatory, an anodyne (pain relieving), and antibacterial. They work wonders on mosquito bites, bee stings, rashes, and minor cuts and scrapes.

There are three ways to use the leaves for first aid. The simplest is to crush up a leaf and rub it on the bite or scrape. You can also turn the leaves into an herbal ointment. But by far the most effective way to use plantain (if you aren't grossed out by it) is to make a spit poultice. Chew one of the leaves for a moment and then apply the wad of chewed-up leaf.

Plantago Chips

Makes 24 chips; recipe can be multiplied

I called these "plantago chips" instead of "plantain chips" because the latter name would be too reminiscent of the banana-like fried plantain that has nothing to do with this recipe. Here, the otherwise stringy veins of Plantago species are transformed into extra crunch in a tasty snack. These chips are all about texture; I have to admit the leaves themselves are fairly bland. But they are a perfectly crisp vehicle for whatever seasoning you put on them.

INGREDIENTS

24 large leaves of any *Plantago* species

2 teaspoons olive oil

¼ teaspoon salt

½ teaspoon seasoning (The Wild Kimchi Powder on page 223 is fabulous here, as is the Wild Garlic Powder on page 252. You could also use cayenne, any seasoned salt, nutritional yeast, or any seasoning that is good on kale chips.)

INSTRUCTIONS

1. Preheat the oven to 250°F.

2. Wash the *Plantago* leaves and dry them well in a salad spinner or by rolling them up in a clean dish towel.

continued

3. In a large bowl, toss the leaves with the oil until they are well coated. Spread the leaves in a single layer on baking sheets. Depending on the size of the leaves you gathered, you may need to use more than one baking sheet or to bake them in batches. They can overlap a little, but it is important not to crowd or stack them.

4. Sprinkle the leaves with the salt and seasoning. Bake until crisp but not burnt, which may take anywhere from 10 to 20 minutes, depending on the size of the leaves. Remember that they will continue to crisp up a bit as they cool, just like cookies do after you take them out of the oven. If you aren't sure if they're done, err on the side of underdone. Take them out, let them cool for just a minute, and put them back in the oven if they're not crunchy enough.

5. Once they are completely cooled, you can store your *Plantago* chips in an airtight container for several weeks. If the container is not airtight, the chips may absorb some humidity from the air and lose their crispness. Not a problem: simply put them back into a 250°F oven for 3 to 5 minutes.

Prickly Pear/Tuna/Sabra/Beaver Tail Cactus
OPUNTIA SPECIES

That's a lot of common names for one cactus genus, but it's because this cold-hardy plant is beloved on several continents. Once you know how to get past its prickly armor, you'll be able to harvest both a sweet fruit and a versatile vegetable.

Find

Prickly pears grow in many regions worldwide. Native to the Americas, they have been introduced to other parts of the world, including Europe, Africa, the Middle East, and Australia. *Opuntia* cacti are cold tolerant and often thrive in places where people don't expect to see cacti growing, like in northeast North America. Look for them in full sun on rocky or sandy soil.

Identify

Opuntia cacti have flat, paddle-shaped stem sections called cladodes. They are not "cactus leaves" as many people think, although they function like them, capturing the energy of the sun. The younger, lighter green cladodes grow on top of the fleshier, dingier, older ones. It's those young paddles that are eaten as the vegetable nopales. The edges and flat surfaces of both young and old paddles have spots of very small, easily detachable spines called glochids. It's these glochids—more than the sometimes present large spines—that can get into your skin and cause discomfort for days.

Prickly pear's cup-shaped flowers are usually yellow, but sometimes pink or white.

The fleshy fruits have numerous common names, including "tunas" and, in Israel, "sabras." They grow directly out of the edges of the paddles. They are oblong, with a round crater at one end. The fruit may be red, orange, yellow, or even green when ripe. The color of the juicy pulp inside also varies from species to species, sometimes orange, sometimes a striking magenta color. Each fruit contains numerous seeds.

Sustainably Harvest

Prickly pears are considered invasive in Australia and in Mediterranean countries. You are definitely not harming the plant by collecting the "tunas" (fruits) or a few of the nopales (paddles).

Harvesting prickly pears can be challenging—they're called "prickly" for a reason. It's not so much the actual spines that you have to watch out for as much as the much smaller, hair-like glochids that are nestled around the bases of the spines. Get these under your skin and you'll be uncomfortable for days (to put it mildly). In my experience, gloves, even heavy work gloves, are not sufficient to protect you from glochids. Here's a way to harvest prickly pear fruit bare-handed without suffering a glochid attack:

Find a small plastic beverage bottle and cut it in half crosswise. Hold the bottom half so that the base end is in the palm of your hand. Slip the open end over a ripe *Opuntia* fruit. Squeeze the plastic around the fruit and give it a twist. It will come off into the plastic bottle. Drop it into a solid container or basket (plastic and cloth bags quickly become useless with this particular harvest). Move on to the next one.

To harvest the paddles, or nopales, look for those at the top of the cacti that are 8 inches or less in length and fairly thin. They will be a much lighter and brighter green than the older, thicker paddles. There should be no paddles growing from them (in other words, you're harvesting the ones that are at the end of their "branch"). There should also be no fruits growing from them. Older, thicker paddles tend to be stringy and flabby—textures most people do not find appealing in food.

Gloves will not protect you sufficiently from this harvest's bite. Instead, here are two ways to gather the nopales. You can simply hold an open, solid container under or alongside the paddle you're after and use scissors or garden pruners to cut it off at the base so that it falls into your container. Or, fold a piece of cardboard in half and use it to surround and hold the paddle while you use your other hand and a tool to cut off the base.

Eat

Obviously, you need to get rid of the spines and pesky glochids before eating prickly pear fruit or nopales. Wearing gloves for this process may or may not protect you, but it can't hurt.

For the fruits, use tongs to hold them over an open flame such as that of a gas stove or barbecue grill. Rotate the fruit, adjusting the tongs several times, so that every part of its surface is exposed to the glochid-burning flame. Rinse the fruit well under cool water and cut it in half lengthwise. Scoop out the sweet flesh, discarding the skin. (Trust me, you don't want to try the scooping part until you've rendered the skin harmless.)

Prickly pear fruit is fabulous as juice, jelly, syrup, sorbet—basically any preparation that eliminates the numerous seeds.

To deal with the seeds, you can boil the pulp and then strain, as in the prickly pear syrup recipe below. Or you can put the pulp in a blender and pulse it a few times before straining through a fine-holed colander. This is the quicker method, but results in cloudy syrup or jelly. It is fine for any other type of recipe.

For nopales, use tongs or a folded piece of cardboard to grasp one side of the paddle while with your other hand you use a sharp paring knife to scrape both flat sides. You want to remove the spines, but keep as much of the nopales' green skin as possible because it improves the texture of the cooked vegetable.

Once you have removed the spines by scraping the flat sides, trim off the edges all the way around. For extra insurance, hold each paddle with tongs and hold it over a flame such as that of a gas stove to burn off any remaining glochids. Rinse the paddle under cold water.

Nopales have a crisp-tender texture when harvested at the young stage I describe above. But they can be slimy if not prepared correctly. I recommend parboiling and rinsing them, even if you will be using another cooking method (deep frying, grilling, stir-frying) afterward. This gets rid of a lot of the sliminess. To do this, first cut them into strips or small pieces. Boil them in salted water for 15 minutes, then drain well. Rinse under cool water for a couple of minutes. If the recipe calls for boiled nopales, you probably won't need to reboil them for more than another 5 minutes.

Prickly Pear Syrup

Makes 2 pints

Prickly pear fruits (sabras) make a delicious and naturally colorful syrup. Of course you can use this syrup on top of pancakes, but it is also wonderful in cocktails or as the base for a granita or sorbet.

INGREDIENTS

12 large or 24 small prickly pear fruits, glochids and spines singed off

2 cups granulated white sugar

3 tablespoons lemon juice

INSTRUCTIONS

1. Scoop the pulp out of the fruits and coarsely chop the pulp. Put it into a large pot along with enough water to cover the fruit. Bring to a boil over high heat, then reduce the heat and simmer until the fruit pulp is losing its color and flavor to the liquid, about 15 minutes.

2. Strain the liquid through a jelly bag, several layers of cheesecloth in a colander, or a muslin bag.

3. Measure the strained liquid. There should be about 4 cups. If not, adjust the amount of sugar so that you use ½ cup sugar per cup of prickly pear juice. Return the prickly pear liquid to the pot (wash the pot first if there are any seeds sticking to it). Add the sugar and lemon juice. Bring the mixture to a boil over medium-high heat, stirring to dissolve the sugar.

   A foamy scum usually forms on the surface of the prickly pear syrup while it is coming to a boil. Skim as much of this off as possible and discard it.

4. Pour the still-hot prickly pear syrup into clean pint or half-pint canning jars (it is not necessary to sterilize the jars for this recipe). Be sure to leave ½ inch of headspace between the surface of the syrup and the rims of the jars. Wipe the rims of the jars clean with a damp paper towel or clean dish cloth.

5. Screw on the canning lids and process in a boiling water bath for 10 minutes (adjust the canning time if you live at a high altitude). See the Useful Resources section if you're unfamiliar with canning.

   If you'd rather skip the canning process, you can refrigerate your prickly pear syrup for up to 1 month or freeze it for up to 1 year.

Nopales Scramble with Salsa Verde

Serves 6

Scrambled eggs are a traditional vehicle for nopales, and the two together have a great, silky texture. The salsa verde brings intense flavor to the party.

NOPALES INGREDIENTS

8 nopales, glochids and any spines removed
1 teaspoon salt
1 tablespoon olive oil
1 medium onion, chopped
6 eggs, beaten

SALSA VERDE INGREDIENTS

2 garlic cloves, peeled and minced or pressed
2 tablespoons capers
4 anchovy fillets, chopped (optional)
1 cup flat-leaf parsley leaves
1 cup fresh basil leaves
1 cup fresh mint leaves
1 tablespoon Dijon mustard
3 tablespoons red wine vinegar
½ cup good-quality extra-virgin olive oil
Salt and freshly ground black pepper to taste

INSTRUCTIONS

1. Cut the nopales into ¼-inch-wide strips. Put the salt in a large pot of water and bring to a boil over high heat. Add the nopales strips and boil for 15 minutes. Drain in a colander, then rinse well under cool water.

2. While the nopales are boiling, prepare the salsa verde. You can finely chop all of the herbs by hand or pulse all of the ingredients together in a blender or food processor.

3. Heat the oil in a large skillet. Chop the nopales strips into ½- to 1-inch pieces and add them to the pan with the onion. Cook over medium heat,

stirring often, until the onions start to caramelize and the nopales are tender and most of their liquid has evaporated.

4. Reduce the heat to low and add the eggs. Cook, stirring constantly, until the eggs are set but still on the soft side. (The texture will be better if you resist the urge to crank the heat.) Spoon the salsa verde over the top or serve it on the side.

Purslane
PORTULACA OLERACEA

One of summer's most ubiquitous weeds, this succulent plant is not only delicious, but also contains healthy omega-3 fatty acids.

Find

Purslane likes hot weather and only appears in late spring when nighttime temperatures finally warm up. Look for it growing in full sun in unweeded gardens and flower beds, sidewalk cracks, parks, new building lots without any grass yet, roadsides, farms, and other disturbed-soil environments.

Identify

Purslane is a low-growing plant with succulent, often reddish stems and small (¼ to 1¼ inch long), oval- or spade-shaped shiny green leaves. Its little yellow flowers have five petals. The oval seed capsules split open around their middles to release the many minuscule black seeds within. Although the plants only grow a few inches high, they can sprawl on the ground to as wide as 2 feet across.

Note that there is a similarly low-growing poisonous plant called spurge that likes the same kind of habitat and often grows alongside purslane. It does not have purslane's succulent, hairless stems, and it exudes a white latex when the stringy stems are broken, which purslane does not.

Sustainably Harvest

Purslane self-seeds so readily that, even though it is cultivated as a desirable crop in some countries, it is considered an invasive weed in others. This is another plant with which you do not have to worry about sustainability.

Collect purslane's leaves and stems anytime during its growing months. The flowers and seeds are edible, too, but too small to spend time on. If a few wind up in your collection bag, just eat them along with the rest of the plant.

Eat

I've seen purslane for sale at farmers markets in France, on Crete, in New York City's greenmarkets, and on the steps of the Old City in Jerusalem. I'm sure there are many other places where it is valued, and it's too bad that most of the time gardeners unknowingly toss this treat into the compost bin.

Purslane is juicy with a sour, lemony taste. The stems and leaves are good raw in salads. Classic purslane recipes often include it in potato salad, or instead of cucumber in a tzatziki-like salad with yogurt.

Cooked, purslane's mucilaginous property comes out. This makes it terrific for thickening soups, stews, and gumbo, although to my tastes less desirable as a cooked green on its own.

The fat stems make excellent pickles and relish.

Purslane Relish

Makes 2 to 3 half-pint jars

Use this sweet-and-sour relish in place of hot dog relish. It's also excellent in deviled eggs or mixed with mayo and served with seafood.

INGREDIENTS

4 cups finely chopped fresh purslane, leaves and thick stems or just stems

1 to 2 medium-large onions, peeled

1 large red or orange bell pepper, stem and seeds removed

2 tablespoons kosher or other non-iodized salt

½ teaspoon celery seed

¼ teaspoon mustard seeds

¼ teaspoon ground mustard

¼ teaspoon turmeric powder

¼ teaspoon ground nutmeg

½ cup apple cider or white wine vinegar

¾ cup sugar

INSTRUCTIONS

1. Strip the leaves off the purslane (save them for another use such as salad). You're just going to use the thick, succulent stems for the relish.

2. Chop the purslane very finely or pulse it a few times in a food processor. Transfer the finely chopped purslane to a large bowl.

3. Cut the onion and bell pepper into big chunks, then finely chop them or pulse them a few times in a food processor. You want the vegetables to be in tiny pieces, but not totally pulverized into a purée. Add the onion and pepper to the purslane in the bowl.

4. Add the salt to the vegetables and mix well. If it seems like a lot of salt, don't worry: You'll be rinsing most of it off later. The salt is there to draw water out of the vegetables, a step that results in better texture and flavor in the finished relish. Cover the bowl of vegetables and leave it in the refrigerator for 8 hours or overnight.

5. Transfer the vegetables to a finely meshed sieve or strainer and let them drain for a couple of minutes. Don't be concerned if the liquid that drains

8. Bring the vinegar and sugar to a boil in a medium-size pot. Once the mixture is boiling, add the vegetables. Let the mixture return to a boil, then immediately turn off the heat.

9. Strain the hot relish in a sieve placed over a bowl to catch the liquid. You'll need the liquid for the next step.

10. Loosely pack the relish into clean, hot canning jars, leaving at least ½ inch of headspace. It is not necessary to sterilize the jars for this recipe. Pour the hot brine from the bowl over the relish; the vegetables should be completely submerged in the brine. Press lightly on the relish with the back of a small spoon to release any air bubbles.

11. Wipe the rims of the jars with a paper towel or clean cloth. Screw on the canning lids. You could simply store the relish in the refrigerator, but for longer storage at room temperature, process in a boiling water bath for 10 minutes (adjust the canning time if you live at a high altitude). See the Useful Resources section if you're unfamiliar with canning.

Wait at least a week for the flavors to develop before eating your purslane relish.

out is somewhat goopy: that same mucilaginous property means we don't have to add any of the usual thickeners such as cornstarch to this relish.

6. Rinse the vegetables thoroughly under cool water and let them drain again. Get even more of the liquid out by pressing the vegetables against the sieve with the back of a wooden spoon.

7. Return the rinsed and drained vegetables to the bowl, add the spices, and stir to combine.

Gumbo with Purslane and Wild Sassafras

Serves 5 or 6

Purslane takes the place of the usual okra in this gumbo. It is included not only for its thickening properties (doesn't that sound better than "mucilaginous"?), but also for its lemony taste that goes beautifully with the seafood.

If you can't find wild sassafras growing where you live, you can mail-order filé powder, which is simply dried, powdered sassafras leaf.

INGREDIENTS

¼ cup vegetable oil

¼ cup all-purpose flour

1 medium onion, peeled and chopped

1 large red bell pepper, stemmed, seeded, and chopped

¼ cup celery stalks and leaves, chopped

1 to 3 fresh or dried hot chili peppers

4 cups chicken or vegetable stock

1 cup tomatoes, chopped (canned is fine)

2 cups purslane leaves and stems, chopped

2 links andouille, chorizo, or other smoked sausage cut into ½-inch chunks

1 garlic clove, minced

1 pound seafood (pieces of whitefish, scallops, shrimp, or a combination of these)

1 tablespoon filé powder

Salt to taste

Cooked rice to serve the gumbo over

INSTRUCTIONS

1. In a Dutch oven or other heavy pot, heat the oil over medium heat. Whisk in the flour. Stir constantly with a wooden spoon until the roux is a medium brown color. Remove from the heat and let cool for 1 minute. This is the roux that is the foundation of many Creole sauces, and it's an absolute requirement for gumbo. Along with the purslane, it will help thicken the dish as well as add rich taste and texture.

2. Add the onion, bell pepper, celery, and hot peppers. Return to heat. Whisk in the stock and tomatoes. Bring to a boil over medium-high heat, whisking frequently.

3. Reduce the heat to medium. Add the purslane, sausage, and garlic. Simmer until the sausage is cooked and the vegetables are tender, 10 to 15 minutes.

4. Add the seafood and cook 5 minutes more.

5. Turn off the heat. Stir in the filé powder and salt to taste. Serve over cooked rice.

VEGAN VARIATION

Leave out the sausage and seafood and add 2 cups cooked beans. Substitute 1 to 2 chipotle peppers in adobo for the hot chili peppers to give the gumbo a smoky flavor.

Quickweed/Guacus/ Gallant Soldier

GALINSOGA PARVIFLORA AND G. CILATE

Often dismissed in field guides as "bland," quickweed is the perfect companion for bitter or strongly flavored wild greens. Considered a weed in North America and Europe, it is cherished as a choice vegetable in Colombia and eastern Africa.

Find

Look for quickweed in partial shade to partially sunny spots in the disturbed soils of parks, gardens, farms, and roadsides.

Identify

Quickweed's somewhat hairy leaves have an ovate shape (broader at the base than the tip) and grow alternately on the also-hairy stems. If you just give the flowers a quick glance, you might think they look like tiny daisies with their yellow centers and white petals. But a closer glance reveals that each little flower has just five petals, and each of those petals has a three-lobed tip.

Note that there is a similar-looking plant, *Tridax procumbens*, which is not edible. Quickweed grows upright, whereas *Tridax* sprawls on the ground. *Tridax* is used in India to treat diabetes.

Sustainably Harvest

Pinch off and use the top few inches of the plants—stems, leaves, flowers, and all. As its common name suggests, this plant grows quickly and is considered

mildly invasive, so there are no sustainability concerns with this one.

Sometimes leaf miner bugs attack quickweed, leaving behind what look like white trails on the leaves. I don't harvest these infested leaves, not because they're dangerous but because they aren't appealing.

Eat

Quickweed's hairy leaves and stems rule it out as a raw salad ingredient in my opinion (although it would be safe to eat raw). When steamed, boiled, or stir-fried, that hairiness disappears and what you are left with is a pleasantly mild cooked green. I like to combine it with greens like wild mustard that have a more assertive taste. Quickweed is good in omelets and quiches. It's also good in strongly seasoned dishes like curries, where it creates a neutral background for the sauce.

Called guascas in Colombia, it is used there in a dish called *ajiaco*.

Ajiaco Guascas (Colombian-Style Chicken and Potato Soup)

Serves 6

This is a satisfying and easy one-pot meal.

INGREDIENTS

¼ cup olive oil

1 large onion, chopped

6 garlic cloves, peeled and chopped

2 large chicken breasts (about 1½ pounds), cut into 1-inch-thick pieces

2 teaspoons sea or kosher salt

4 cups chicken stock

1½ pounds potatoes, peeled and cut into 1-inch pieces

3 ears fresh corn, cut in half crosswise

1 cup scallions, sliced

1 large bunch coriander leaves (cilantro), washed and with the stem end tied securely with kitchen string

⅓ cup dried or 1 cup fresh guascas (quickweed leaves)

TOPPINGS

2 avocados, pitted, peeled, and cut into thin slices

½ cup sour cream

½ cup coriander leaves (cilantro), chopped

2 tablespoons capers

INSTRUCTIONS

1. Heat the olive oil over medium heat in a heavy-bottomed pot such as a Dutch oven. Add the onion and cook, stirring often, for 5 minutes. Add the garlic and cook, stirring constantly, for 1 minute more.

2. Add the chicken and cook for 5 minutes.

3. Add the salt, stock, and potatoes and bring to a boil over high heat. Reduce the heat to medium-low and simmer for 10 minutes.

4. Add the corn, scallions, cilantro, and quickweed. Cook for 20 minutes more.

5. Remove the bunch of cilantro. Serve the soup with any or all of the toppings.

Redbud
CERCIS CANADENSIS (EASTERN NORTH AMERICA), C. OCCIDENTALIS (WESTERN NORTH AMERICA), AND C. SILIQUASTRUM (MEDITERRANEAN REGION AND AUSTRALIA)

Redbud's bright pink blossoms are one of the glories of spring, but they are not just eye candy. These bright blossoms taste like lemony green beans.

Find
Look for redbuds on sunny slopes or occasionally as an understory tree or shrub. It is common in urban and suburban areas, where it is a favorite with landscapers.

Identify
Redbuds are small trees with branches that often grow in a quirky zigzag pattern. The bark of young redbuds is smooth, but becomes craggy as the trees mature. The clusters of pink or magenta pea flower–shaped blossoms seem to be growing directly out of the bark, appearing before the heart-shaped leaves. Those leaves, which have smooth edges and stalks with slightly swollen bases, grow in an alternate pattern on the branches.

In late spring and early summer, the flowers become pods that look like small snow peas. Once brown and mature, these seedpods often persist on the branches long after the trees have dropped their leaves in autumn.

Sustainably Harvest
Collecting the blossoms doesn't damage the trees, but it is never a good idea to strip the branches bare. Leave plenty of flowers on each branch so that others can enjoy their beauty and so that they can mature and go to seed.

Eat
Besides adding gorgeous color to food, redbud blossoms have an interesting flavor that starts out with a green bean–like taste and then develops into a pleasantly sour aftertaste. They are fantastic raw on salads, but can also be pickled, added to sorbets and ice creams, and baked into muffins and baked goods.

The very young pods can be cooked like snow peas.

Bloom 'n' Lentils

Serves 4 as a side dish, 2 as a main course

In this dish the lentils match the redbud flowers' beany taste, while the sorrel brings out their tangy second flavor.

INGREDIENTS

1 cup small black or green lentils

2 to 3 cups water

¼ cup extra-virgin olive oil

Juice of ½ lemon

½ teaspoon salt

¼ teaspoon dried or ¾ teaspoon fresh thyme leaves

¼ teaspoon freshly ground black pepper

1 cup redbud blossoms

¼ cup wood sorrel leaves and flowers; sheep sorrel leaves; or garden sorrel

¼ cup wild garlic flowers, separated from their clusters; minced wild garlic leaves; or minced chives

INSTRUCTIONS

1. Put the lentils and 2 cups of the water into a small pot. Bring to a boil, then reduce the heat and simmer until the lentils are tender. Add more water if necessary to keep the pot from drying out before the lentils are cooked, which will take 20 to 30 minutes.

2. While the lentils are cooking, whisk together the oil, lemon, and seasonings. When the lentils are ready, drain them in a colander, then immediately transfer them to a bowl. While they are still warm, stir in the salad dressing. Let the lentils cool to room temperature before proceeding to the next step.

3. Gently stir the redbud, sorrel, and wild garlic or chives into the lentils, keeping a few of the flowers aside to sprinkle on top as a garnish. This salad is best served at room temperature and eaten immediately or within a few hours of making it.

Pickled Redbud

Use these piquant and colorful pickles instead of capers in any recipe that calls for the latter. The texture of these pickled redbud blossoms is best if you collect the flower buds before they have fully opened.

INGREDIENTS

Redbud blossoms
White wine or distilled vinegar
Water
Kosher or other non-iodized salt

INSTRUCTIONS

1. Rinse the flower clusters under cold water. Pinch off and discard the stems.

2. Combine equal parts white vinegar and water. Add ½ teaspoon salt per cup of brine and stir to dissolve. Plan on an equal amount of brine by volume for the quantity of redbud blossoms you have gathered. In other words, 1 cup of brine per cup of flowers.

3. Fill a clean jar with the redbud blossoms, then cover them with the brine. Make sure the jar is completely full, then simply screw on the lid to keep the blossoms submerged under the brine. Some brine will leak out when you do so; that's okay. Place the jar on a small plate and leave at room temperature for 3 days, away from direct sunlight which could discolor the flowers.

4. Transfer the jar to the refrigerator or a cool cellar. Don't expose pickled redbud blossoms to heat or their texture and color will diminish.

Rose

ROSA SPECIES

Roses aren't just a blessing for our senses of sight and smell. Their petals can be delicious, their fruits are packed with vitamin C, and their leaves make a pleasant tea.

Find

Of course there are plenty of roses deliberately planted in gardens, but you'll also find the plants growing wild. Look for them in full to partial sun. Some species are salt-tolerant and are frequently found near beaches.

Identify

Roses grow on stems called canes that are infamously thorny. But here's a botanical fact for you: Roses don't technically have thorns, they have prickles. The difference is that real thorns emerge from the wood of the plant (think hawthorn), whereas prickles come from the outer layer of the stems and break off easily.

Rose leaves each have an odd number of leaflets with toothed edges. They attach to the stems in an alternate pattern.

The renowned flowers come in colors from burgundy so dark it's almost black to red, pink, lavender, yellow, and white. Pink and white are the most common colors for wild roses. What they all have in common are numerous yellow stamens (the pollen-bearing parts) at the center of the flowers. This is a hallmark of plants in the Rosaceae family, which

includes blackberries and other brambleberries as well as roses and many other plants. Also typical are pointy, star-shaped green calyxes at the base of the flowers, which usually remain to clasp the bottoms of rose hips, as rose plant fruits are called. Wild roses typically have a single ring of petals, unlike the abundance of tightly overlapping petals in horticulturally bred roses.

Rose fruits, or hips, range from as small as ¼ inch in diameter to more than an inch across. They usually have a five-pointed crown on one end and small hairs on the skins of the fruits. Rose hips frequently cling to the prickly canes long after the plants have dropped their leaves for the winter. It is worth learning to identify the plants during that leafless, flowerless stage because that is the best time to harvest the hips.

Sustainably Harvest

Before you harvest the flowers, smell them. Remember that there is a direct correlation between the strength of their fragrance and the potency of their flavor. Assuming you've found aromatic roses, it's best to harvest them by snipping them off a little way down the stem just above a leaf. You can remove the petals for use when you get home. But if you snap off just the flower, all of the stem between the cut and the next node below will turn brown. This isn't dangerous for the plant, but it's not pretty for the next human who comes by to smell the roses.

The hips can be harvested any time after they turn orange or red (depending on the type of rose). However, their flavor is better after they've survived a few frosts in late fall or early winter. They may have shriveled up slightly by then, but they will be tasty.

You are not harming the plants by harvesting the flowers or fruits (hips). With the leaves, you need to be a bit more cautious. Never take more than 20 percent of the leaves from a plant, and be careful not to yank them off in a way that strips the skins of the canes (rose leaves usually break off cleanly, so this isn't a big problem). By late summer, rose leaves often suffer from blackspot, a fungal disease, so it is best to harvest them for tea earlier in the season.

Rose growers often treat their roses with a fungicide, so keep this in mind if collecting where the roses are not wild.

Eat

Rose petals and the rosewater made from them are used in many North African, eastern European, and Middle Eastern dishes. You can use the petals fresh, or dry and powder them to make a wonderful spice. (Try mixing powdered fragrant rose petals with cinnamon and adding that to hot chocolate for something heavenly.)

Rose hips can be used fresh, dried, or frozen. You'll need to remove the pointed green calyx at the base, and also cut them in half and scrape out the hairy centers. I skip that last step with really small wild rose hips and simply dry them whole for use in tea blends. You can also skip it for recipes that will be strained, such as wine, syrup, or jelly. But for jam it is essential to get rid of that central "choke," with the irritating hairs.

Rose leaves make a mild but interesting tea that is best made with dried rather than fresh leaves. During the rationing of World War II, when black tea was often unavailable, people resorted to rose leaf tea. It doesn't have the caffeine of black tea, but it does have quite a few tannins, which give it a similar mouthfeel. The flavor is very mild, and I like to blend it with other tea herbs such as mint or pineappleweed.

Rose Petal Honey

Makes 1 cup

In Greece you can buy tins of rose petal jam. This mellite is thinner than that jam, but with a similar flavor. It is not honey made by bees that pollinated rose plants, but rather honey mixed with actual rose petals. It is absolutely essential that the roses you use are extremely fragrant. Fortunately, many wild roses are more aromatic than their cultivated counterparts.

INGREDIENTS

1 cup fresh, fragrant rose petals

1 cup light honey (clover or wildflower honey works well; you want something that doesn't have an overpowering flavor of its own)

¼ teaspoon rosewater (optional)

INSTRUCTIONS

1. Finely chop the rose petals and put them into a clean glass jar.

2. Pour the honey and rosewater (if using) over the rose petals. Stir and let sit uncovered for 5 minutes while the honey settles into any air pockets. Cover and store at room temperature away from direct light or heat. Wait at least 2 weeks for the roses to give up their flavor to the honey before tasting. The honey will liquefy a bit from the moisture in the petals, but this isn't a problem. Keeps indefinitely.

Try adding rose petal honey to iced tea, or drizzle it over fresh fruit or ice cream.

Rose Hip Freezer Jam

Makes 2 pints

Recipes that require cooking rose hips to turn them into spreadable jam destroy their naturally high vitamin C content. This freezer jam is made with raw rose hips, so most of the vitamin content is preserved. Unfortunately, you have to use a lot of sugar for the powdered pectin to work. If you want a low-sugar version, use a low-methoxyl pectin such as Pomona's Universal Pectin.

INGREDIENTS

½ pound ripe rose hips

1½ tablespoons lemon juice

1½ cups water

2 cups sugar

1 package powdered pectin

INSTRUCTIONS

1. Wash the rose hips and slice off the "crown" end. Cut them in half. Next comes the most labor-intensive part of making rose hip jam: Scoop out and compost or discard the hairy, seedy centers. A serrated grapefruit spoon is a useful tool for this.

2. Put the prepared rose hips, lemon juice, and ¾ cup of the water into a blender or food processor.

3. Purée the mixture until it is fairly smooth (a few bits of rose hip skin are okay).

4. With the blender or food processor running, add the sugar a little at a time. When the sugar is incorporated into the other ingredients, turn off the machine. If the sugar has not completely dissolved, let the fruit mixture sit at room temperature for 30 minutes to an hour before proceeding.

5. In a small saucepan, stir together the pectin and another ¾ cup water. Bring to a boil over high heat. Continue to boil for 1 minute.

6. Pour the pectin liquid over the other ingredients in the blender or food processor. Blend for about 30 seconds until everything is well combined.

7. Transfer the rose hip jam into small freezer containers, leaving ½-inch headspace between the top surface of the jam and the rims of the containers. Cover and store in the freezer for up to a year.

8. To use your rose hip freezer jam, take out one of the containers and let it thaw in the refrigerator. Once you've moved it from the freezer to the refrigerator, use your rose hip jam within 1 month.

Sea Lettuce/Green Laver
ULVA SPECIES

Emerald green and found on coastlines around the world, sea lettuce is packed with nutrients. It makes a tasty seasoning, is good in soups and salads, and can also be used as a nori substitute when making sushi.

Find

Sea lettuce grows in cool waters on intertidal rocks and tide pools around the world. It also thrives in freshwater runoff such as the mouths of rivers and streams, and sewage pipe runoff, which is most definitely *not* a recommended place to forage. It frequently attaches to wooden jetties, driftwood, and shells.

All of the different species in the *Ulva* genus are edible.

Identify

Look for seaweed that is a much lighter, brighter green than the other seaweeds that may be growing near it. In fact, sea lettuce is such a vivid light green that it is easy to spot from a distance.

The "leaves" (technically the blade or thallus) are only two cells thick, which gives them a translucent, delicate appearance.

Sustainably Harvest

Sea lettuce is stronger than its almost transparent look would lead you to believe, but it is still easy to harvest using a sharp knife or scissors. Grab a handful of it with one hand while you cut it off just above where it is attached to the rock, wood, or whatever it's clinging to.

Eat

Add raw sea lettuce to soups and salads. Keep in mind that it has a strong, briny taste—a little goes a long way.

When dried, sea lettuce loses some of its emerald color and darkens, but retains its flavor and all its wonderful nutrients. Sea lettuce is high in iron and calcium, and also contains protein, iodine, vitamin B1, and vitamin C, among other nutrients. It can be ground into flakes or a powder to use as a seasoning that is good on rice, seafood, tofu, and steamed vegetables.

Nori, the paper-like sheets of seaweed used to

wrap sushi, is usually made with seaweeds from the genus *Porphyra*. However, sea lettuce also has been used. Making nori is a lot like making paper. First you blend the seaweed into a pulp with a little water. Then you spread the pulp out very thinly and dehydrate or bake it until completely dry. It is completely possible to make your own nori at home on the lined sheets of a dehydrator set to 110°F.

Green Laver Gomasio; recipe follows

Green Laver Gomasio

Makes 2 cups

Gomasio is sometimes made with just sesame seeds and salt; other times with the addition of seaweed. People following a macrobiotic diet use it on brown rice and make all sorts of health claims for it. They are right about it being tasty on rice, but I sprinkle it over many other foods as well.

INGREDIENTS

1½ cups unhulled sesame seeds
½ cup dried green laver (sea lettuce) flakes
3 tablespoons fine-grain sea salt

INSTRUCTIONS

1. Heat a cast-iron or other heavy-bottomed skillet over medium heat. Add the sesame seeds and cook, stirring constantly, until they start to make popping sounds, become noticeably aromatic, and turn brown. Keep an eye on them, because the difference between nicely toasted and burnt can be a matter of moments. Remove the seeds from the skillet and set them aside.

2. Put the skillet back over the heat and add the sea lettuce. Cook, stirring constantly, for 1 minute. Add the toasted sea lettuce to the sesame seeds. Return the skillet to the heat.

3. Put the salt in the skillet and cook, stirring, for 2 minutes. If you are using a cast-iron skillet, the salt will turn a grayish color. Add the salt to the other ingredients.

4. Grind the gomasio with a mortar and pestle or by pulsing it in a food processor or electric grinder. You don't want to completely reduce it to a powder. There should still be recognizable flecks of sea lettuce and specks of sesame seeds, but the sesame seeds should be at least 50 percent ground.

   Store gomasio in a tightly sealed jar in the refrigerator or other cool place for up to 1 year. (If you store it at a warm room temperature, the sesame seeds may turn rancid.)

Sheep Sorrel
RUMEX ACETOSELLA

With the same delightfully sour flavor as its cultivated cousin, garden or French sorrel, sheep sorrel can be used raw or cooked.

Find

Sheep sorrel loves sunny lawns and other bright, open areas. Although it likes to grow in sun, the leaves are bigger and more tender in locations where the plants get a few hours of shade.

Identify

Sheep sorrel leaves have the same arrowhead shape as cultivated sorrel, but they are much smaller, usually only 1 to 3 inches long. There is a rosette of leaves at the base (that's where most of the leaves are), and also a few alternately arranged leaves on the flower stalks. These flower stalks can grow up to 18 inches tall. It's common for the leaves on the flower stems to lack the arrowhead shape of the rosette leaves and instead be simply linear. There is a thin sheath where the leaf stalks attach to the stems. Sheep sorrel's flowers and seeds are tiny, yellowish to rusty red, and grow in clusters.

Sustainably Harvest

Sheep sorrel is in season from spring through fall. It's best to harvest from the rosette leaves before the plants start to flower because the flower stems are stringy. You can extend the harvest of tender rosette leaves by keeping the flower stalks pinched off.

To harvest sheep sorrel, gather a bunch of it together with one hand, as if you were holding a bouquet. With the other hand, cut or twist off the leaves. Discard any tough flower stalks.

Eat

Use *Rumex acetosella* raw or cooked in any recipe that calls for French sorrel (*Rumex acetosa* and *R. scutatus*). You also can use it in place of wood sorrel

(*Oxalis* species; no relation but the taste is pretty much identical). It is terrific paired with seafood.

When you cook sheep sorrel it loses the bright green color of the raw leaves and turns a dull khaki color. As far as I know there's no way to prevent this discoloration when you cook sorrel, but it has no effect on the flavor.

Like spinach, sorrel contains oxalates. In moderate amounts oxalates are not a health concern, but you shouldn't eat large portions of high-oxalate plants every single day, especially if you have a history of kidney stones.

Salmon with Sorrel Sauce

Serves 4

Sometimes there's no sense in trying to reinvent the wheel: French sorrel sauce is a classic, especially when paired with salmon. But you can also use this sauce on poultry, other seafood, or even roasted cauliflower.

INGREDIENTS

4 salmon steaks, approximately 1 inch thick, skins on

¾ cup heavy cream

2 tablespoons butter

1½ cups finely chopped sheep sorrel leaves

1 to 2 tablespoons white wine or vermouth

Salt and pepper to taste

INSTRUCTIONS

1. Preheat the broiler of your oven or toaster oven. Line a baking sheet with parchment paper. Place the salmon pieces on the sheet skin-side down.

2. Broil for 6 minutes. Flip the salmon over and broil for 4 minutes more.

3. While the salmon is broiling, put the cream into a small pot over medium heat. Bring to a simmer. This will keep it from curdling when it meets the acidity of the sorrel.

continued

4. While the cream is heating, melt the butter in a skillet over medium-low heat. Add the sorrel and cook, stirring often, until all of it changes from bright green to a drab olive green.

5. Add the cream to the wilted sorrel and let the sauce come to a simmer. Add 1 tablespoon of the wine or vermouth, 2 tablespoons if you want the sauce to be thinner. Remove from the heat and add salt and pepper to taste (or peppergrass, which is lovely in this sauce).

6. Place one piece of salmon on each plate. Pour the sorrel sauce over the salmon. Serve immediately.

TIP:

If need be, you can make the sauce up to an hour ahead of the salmon. Keep it warm and covered in a low oven (200°F) until the rest of the meal is ready.

Other plants that are great in this recipe: Japanese knot-weed, wood sorrel.

Shepherd's Purse
CAPSELLA BURSA-PASTORIS

Often ignored even by experienced foragers, shepherd's purse is a spicy wild green that is good raw, cooked, or fermented into a kimchi-style condiment.

Find

You'll find shepherd's purse growing in full sun in the disturbed soils of human habitation. Look for it in parks, unweeded gardens and fields, roadsides, and the pavement cracks and tree pits of city streets.

Identify

Shepherd's purse is an annual plant that starts out with a rosette of 2- to 4-inch, deeply lobed leaves with toothed margins. The tiny, four-petaled white flowers bloom on flower stalks that can grow anywhere from a few inches to 2 feet tall, depending on the site. The flower stalks are often leafless, but sometimes have a few leaves that are smaller than the rosette leaves. These upper leaves have bases that clasp the stalks.

The seed capsules are the easiest part of this plant to identify. They are small, flat, triangular to heart-shaped pods up to ⅓ inch across, each holding several reddish-brown seeds. Supposedly the shape of the seedpod matches the purses that shepherds used in centuries past, and that's how shepherd's purse got its name.

The taproots are off-white.

Sustainably Harvest

This is a common weed and there is no sustainability issue with regard to harvesting it.

Collect the leaves from early spring up until the plants start to flower. The flower stalks are excellent when they are just starting to shoot up, before the plants set seed. Use the flowering tops and seedpods anytime while the pods are still green.

Eat

Shepherd's purse can be eaten raw or cooked. I think its mildly spicy leaves and shoots are best combined with other wild greens. Its flowers and immature seedpods make good salad garnishes. The whole plant can be fermented and then dried into a spice as in the recipe below.

Simply dried, the seedpods make a decent spice, but they lose a lot of their flavor after a couple months in storage.

Wild Kimchi Powder

Makes ¼ cup powdered spice

You can use this kimchi in its undried, leafy form, and it is excellent. But the spice that results from first making this spicy, lacto-fermented condiment and then dehydrating and grinding it to a powder is wildly good.

INGREDIENTS

½ cup shepherd's purse leaves, seedpods, and flowers

½ cup wild mustard leaves and seeds, or peppergrass, or garlic mustard (or all three)

1 teaspoon wild or cultivated garlic, chopped

1 to 3 hot chili peppers

1 teaspoon fish sauce or soy sauce

1 teaspoon non-iodized salt

1 cup nonchlorinated water (filtered is fine)

1 big piece cabbage, horseradish, or grape leaf

INSTRUCTIONS

1. Put the shepherd's purse, mustard or other spicy plants, garlic, and chili pepper(s) into a small, clean glass jar.

2. Stir the fish sauce, salt, and water in a separate container until the salt is completely dissolved. Non-iodized salt is used because the iodine that is added to some table salts could discolor your ferment (kimchi is a fermented product). Nonchlorinated water is used because chlorine can kill the healthy, probiotic bacteria needed for a successful fermentation. Because most municipal water is chlorinated, it's a good idea to filter yours before proceeding with this or any other fermentation recipe. If you don't own a filter, you can let the tap water sit for a day so it can outgas the chlorine.

3. Pour the liquid over the other ingredients in the jar. Tuck the cabbage, horseradish, or grape leaf over the top so that it keeps the solids immersed in the brine. Loosely screw on a lid.

4. Place the jar on a small plate and leave at room temperature for 3 to 7 days. During that time, each day remove the top leaf, press down on the

continued

kimchi with the back of a clean spoon to reimmerse the vegetables in the brine, and check for signs of fermentation. You should see some bubbles or foaming action when you press down with the spoon, and the brine should start to have a sour, sauerkraut-like tang to its smell. (Because this is wild kimchi, it also will be quite spicy and pungent.)

The plate under the jar is to protect your countertop from the almost inevitable overflow that will happen during the initial fermentation.

5. Transfer the jar to the refrigerator. It is not necessary to keep a plate under it once you've moved it to the refrigerator. The cool temperature inside the refrigerator will slow down but not halt the fermentation. The longer you leave the kimchi before drying it, the stronger the flavor will be, but wait a minimum of 1 month.

6. Strain the wild kimchi through a fine mesh sieve (you can save the liquid to use as a seasoning in its own right). Spread the remaining solid ingredients on a lined dehydrator sheet. Dry at no higher than 118°F until the ingredients are easy to crumble with your fingers. Alternatively, dry on a baking sheet in an oven on its lowest setting with the door propped open with the handle of a wooden spoon or a dish towel.

7. Grind the dehydrated kimchi in an electric grinder or with a mortar and pestle. Store in a tightly covered glass jar away from direct light or heat.

I enjoy my wild kimchi powder on everything from stir-fried vegetables to hard-boiled eggs to shrimp to popcorn to . . . well, I guess I'm still finding out just how many things I enjoy it on.

Shiso
PERILLA FRUTESCENS

Shiso has a flavor like a cross between coriander and mint, and is traditionally served with sushi and sashimi.

Find

Shiso usually grows wild—or maybe feral would be more accurate—as a garden and landscaping escape. It self-seeds freely and has found its way into city parks, fencerows, and other places close to where it was originally planted. It thrives in both full and partial sunlight, but the purple-leaved form will revert to green if it grows in partial shade.

Identify

There are two varieties of shiso. One has purple-red leaves, and the other has green leaves. Both have square stems like all plants in the mint family; roll one of the stems between a thumb and forefinger and you'll feel the four distinct sides.

Also like other plants in the mint family, shiso has opposite leaves with toothed margins. The leaves are 1 to 5 inches long, rounded at the base and pointed at the tips. They have slender petioles (leafstalks).

The plants eventually grow to be between 1 and 2 feet tall and sometimes branch. The flower heads develop at the ends of the stems and are shaped like elongated bottlebrushes. The small, tubular, individual flowers are purple to white.

The entire plant has a mint-coriander fragrance when crushed, and the scent is your confirmation that you're looking at shiso and not a similarly shaped plant such as coleus, which has no scent.

Sustainably Harvest

Shiso is an annual plant that sets seed in late summer and then dies. Although it self-seeds prolifically, it doesn't really become invasive. I always leave a few of the plants alone so that they can flower and go to seed. When harvesting the others, I pinch off the top few inches of the stems with leaves attached. This encourages new, lush growth, whereas if you pinch off only the bottom leaves, those never regrow and you end up with straggly looking plants and less to harvest in future weeks.

Eat

The flavor of both the green- and the purple-leaved shiso plants is identical. It is a strong taste but a versatile one when used in small quantities as an herb rather than as a main ingredient vegetable.

Shiso is best raw. Use the leaves in sushi rolls or serve them on the side of sushi and sashimi as a palate cleanser. Add it to salads and whole grains. Use the purple-leaved variety to add both color and flavor to pickled ginger and infused sake.

Pickled Ginger (*Gari*) with Shiso

Makes ¾ cup, about 10 servings

Pickled ginger, called *gari* in Japan, is served as a palate cleanser alongside sushi and sashimi. Commercial brands of *gari* come in two colors: the natural light tan of raw ginger root and a bright pink that nowadays usually comes from food coloring. What that pink is *supposed* to come from is very young ginger, which turns ballerina pink when pickled. This recipe gives you that color naturally, even when you're working with older ginger root, and subtly flavors the pickle as well.

INGREDIENTS

4 ounces fresh ginger root, peeled

1 teaspoon non-iodized salt such as kosher or sea salt

4 to 5 large purple shiso leaves, torn into a few pieces each

½ cup rice vinegar

3 tablespoons sugar

2 tablespoons water

INSTRUCTIONS

1. Use a vegetable peeler to scrape off thin slices of the ginger. Place these in a bowl and rub them with the salt until the salt starts to dissolve and lose its gritty feel. Let sit at room temperature for 3 to 4 hours.

2. Transfer the ginger to a sieve and rinse it under cold water. Squeeze out as much liquid as possible, then put the ginger into a clean glass jar. Tuck the shiso leaves in among the ginger (a chopstick helps with this maneuver).

3. Put the vinegar, sugar, and water into a small pot and bring to a boil over high heat, stirring to dissolve the sugar.

4. Pour the brine over the shiso and ginger. Use the back of a spoon to press out any air bubbles and make sure that the brine completely covers the solid ingredients. Cover the jar and refrigerate. At first the ginger will resist the color seeping into the brine from the shiso leaves. Wait at least 1 week for the flavor and color to develop before tasting.

Electric Sake

Makes 1 pint

I first saw and tasted this crazy, wonderful drink at a bar in Tokyo. I was sure they'd spiked my sake with food coloring, because surely that color couldn't have come from anything natural. But it does, and the same purple shiso leaves that give the drink its intense magenta color also add subtle herbal notes to the rice wine.

INGREDIENTS

1 pint sake rice wine

1 cup minced purple shiso leaves, lightly packed

INSTRUCTIONS

1. In a small pot, warm the sake until it is just starting to steam. Do not let it boil.

2. Put the minced shiso leaves into a clean glass jar. Pour the warm sake over the shiso. Cover the jar and let it sit at room temperature for 24 hours.

3. Strain the sake through a fine mesh sieve to remove the spent bits of shiso leaf. Pour the shiso-infused sake through a funnel into a clean glass bottle. Cap or cork and store in the refrigerator or a cool cellar. Serve ice cold.

Sow Thistle
SONCHUS SPECIES

A common weed on several continents, sow thistle has tasty leaves, stalks, and flower buds that can be enjoyed raw, cooked, or pickled.

Find

All of the sow thistle species like disturbed soils (by now you know that means areas where humans have been messing with things: gardens, parks, farms, roadsides, lots . . .) and sunny locations.

Identify

Sow thistles are often confused with wild lettuce and dandelion. Like wild lettuces and dandelions, sow thistle plants exude a white sap when you break off any part of the plant. Sow thistles have a basal rosette of leaves like those other plants, and their yellow flowers are very similar to those of dandelion and wild lettuce. But sow thistle flower stalks are branched and have alternate leaves (dandelion flower stalks are unbranched and leafless). That's true of wild lettuces, too, but sow thistles do *not* have the single line of spines or hairs on the underside of the leaf midribs that wild lettuces have. There are different leaf shapes among sow thistles, but all of them have leaves that clasp the stalks with curved leaf bases.

So here's your checklist (if *all* of these are true, then you've found a sow thistle):

· Basal rosette of leaves

· Flowers like small dandelions

· Whole plant exudes a milky white sap when broken

· Flower stalks are branched and have leaves

· Leaves do *not* have a single line of hairs or spines on the undersides of their midribs

· Bases of the flower stalk leaves curve and clasp the stems

The three most widespread sow thistle species all have those characteristics. Here's how they differ:

Sonchus arvensis (field sow thistle) has much larger flowers than the other sow thistles. It can get as big as 2 inches in diameter. Its narrow leaves are barely lobed at all. Unlike its kin, field sow thistle spreads by horizontal rhizomes and therefore forms colonies.

Sonchus oleraceus (common sow thistle) has a thicker stalk than *S. arvensis*, but much smaller flowers (usually no more than an inch wide). It has deeply lobed leaves.

Sonchus asper (prickly sow thistle) has leaves with wavy edges that may or may not be lobed. Those leaves have much more noticeable prickles along their edges than other sow thistles. The auricles (the curvy part where the leaves clasp the stems) are large and so rounded that those leaves are shaped like commas.

Sustainably Harvest

These are invasive species and sustainability is not an issue.

Always nibble a bit of a sow thistle plant before you decide whether it is worth harvesting. They can be mild, bitter, or even vaguely sweet depending on the maturity of the plant and the location (both soil composition and the amount of sunlight seem to affect the flavor of the plants).

Harvest the leaves and shoots before the plants branch or flower. Pick the leaves off by hand, and break the young shoots off near the base. Pick the flower buds while they are still completely green and closed.

Eat

Use sow thistle leaves raw or cooked. If you've collected prickly sow thistle leaves, you might want to use scissors to cut off the prickly outer edge.

My favorite part of sow thistle is the young flower stalk. Try a nibble raw to decide whether or not the batch you just picked needs to be peeled to be really excellent (peeling will improve the texture and reduce any bitterness, but it often isn't necessary). You can pickle sow thistle stalks or blanch and freeze them for future use.

The green flower buds are good sautéed with other vegetables or pickled.

Vietnamese-Style Spring Rolls with Sow Thistle and Two Sauces

Makes 10 rolls

Visually gorgeous and absolutely delicious, these rolls should be made just before serving. (If you wait longer than half an hour, the rice wrappers become gummy.) The sauces can be made up to a day ahead. Prep all your ingredients before you roll. I like to set all the ingredients on the table and teach guests how to make their own—it's always a lot of fun for everyone.

INGREDIENTS

For the peanut sauce, whisk together:

⅓ cup peanut butter

2 tablespoons soy sauce

2 tablespoons water

1 tablespoon dark (toasted) sesame oil

1 teaspoon rice vinegar

1 teaspoon fresh ginger, grated

1 small garlic clove, minced

If it seems too thick, whisk in a little more water.

For the sweet and sour sauce, purée in a blender:

½ cup jam (plum or peach works especially well here)

3 tablespoons rice vinegar

1 teaspoon soy sauce

1 small garlic clove, peeled

1 small chili pepper, stemmed (leave the seeds if you love spicy; otherwise remove them)

For the rolls:

2 ounces rice vermicelli or bean thread noodles

10 rice wrappers (8½-inch diameter)

2 cups sow thistle leaves and young flower stalks (peeling the stalks is recommended, but not essential), chopped

1 cup cooked chicken, chopped; cooked and peeled small shrimp; or cubed tofu

4 radishes, julienned

¼ cup peanuts, crushed

¼ cup fresh cilantro (coriander leaves), chopped

3 tablespoons fresh mint, chopped

1½ tablespoons Thai basil (optional but really good if you can get it), minced

continued

INSTRUCTIONS

1. Bring a medium-size pot of water to a boil. Cook the rice or bean noodles according to the package instructions (some call for a brief boil, some just for a soak in hot water). Drain, rinse with cool water, drain again.

2. Choose a large bowl with a diameter wider than that of the rice wrappers. Fill it with hot water. Dip one rice wrapper in it for just a second or two (it will continue to soften up after you take it out of the water). Lay it on your work surface.

3. Start piling the ingredients between the bottom and the center of the rice wrapper, beginning with the sow thistle. Add just a little of each ingredient and beware the temptation to overfill, which will result in the wrapper breaking when you try to roll it. Leave at least 1½ inches around the filling ingredients.

4. Fold the bottom of the wrapper over the filling, then the sides, then roll the whole thing up toward the top edge. The wrapper will stick to itself, sealing up in a neat, transparent roll. Or it should. But if the water is too hot or too cool, the wrapper will tend to wad up and be impossible to work with.

That can happen if you're nearing the last rolls and the water has cooled considerably. Keep some hot water handy to warm up the water in the bowl if necessary.

5. Serve at once with either or both of the sauces.

Other plants you can use in this recipe: Basically, any leafy green that's good raw, but Asiatic dayflower is especially good here. You can also throw in some tender but spicy green garlic mustard seedpods or some peppergrass seeds. Shiso leaves are a fantastic addition.

Pickled Sow Thistle Buds

Makes 1 cup

Although these look similar to capers, they have a different flavor.
I dislike the use of the word "capers" applied to anything other than true capers made from the Mediterranean plant *Capparis spinosa*. However, I have to admit that you can use these pickled sow thistle buds in any recipe that calls for . . . well, capers.

INGREDIENTS

1 cup sow thistle buds

1 large garlic clove, peeled and lightly crushed

½ cup apple cider vinegar

¼ cup water

1½ teaspoons non-iodized salt

4 whole black peppercorns

1 bay leaf

continued

INSTRUCTIONS

1. Pack the sow thistle buds and the garlic into a clean glass jar.

2. Put the apple cider vinegar, water, salt, peppercorns, and the bay leaf into a small pot and bring to a boil over high heat. Pour over the sow thistle buds, leaving ½ inch headspace if you will be canning the jar. For a refrigerator pickle version, it's fine to fill the jar all the way.

3. Cover and either store in the refrigerator or screw on a canning lid and process in a boiling water bath for 10 minutes (adjust the canning time if you live at a high altitude). Either way, wait at least a week for the flavor to develop before tasting. See Resources section if you are unfamiliar with canning.

Spruce
PICEA SPECIES

The new growth on evergreen spruce trees is an aromatic ingredient that brings an intriguing and refreshing flavor to a variety of foods and drinks, including seasoning salt, beer, syrup, pickles, and frozen desserts.

Find

Spruce trees grow in boreal forests throughout the Northern Hemisphere. "Boreal" translates to evergreen, coniferous forests that grow across the northern areas of North America and Eurasia. But spruces are also favorites with landscapers and somewhat tolerant of a variety of climates, so you might see them slightly south of their natural range.

Identify

Spruce trees are short-needled evergreen trees. Like pines, they are conifers. That means they have seed-bearing cones. Unlike pines, spruce needles attach singly to their twigs rather than in bundled groups. This is also true of firs, but fir needles are flat, whereas the stout, almost square needles of spruce roll easily between your thumb and forefinger. Another evergreen genus with short, flat needles is yew (*Taxus*). Its needles are poisonous. Remember the rule: If in doubt, toss it out. Although fir's flat needles are edible, if you want to confirm that you have spruce rather than either edible fir or toxic yew, roll the leaves between your fingers and go for spruce's round to squarish needles.

Sustainably Harvest

Collect spruce in the spring when the emerald-hued new growth tips are easy to distinguish from the darker, more mature growth. Pinch off a tip here, a tip there, gleaning from all around the tree rather than wiping out all the new growth on any single branch.

Eat

It is important to use only the "candles"—the light green new growth at the tips of the branches—for spruce recipes. The older growth won't harm you, but it lacks the appealing color and fragrance of the new growth. The older needles are usually too tough and bitter to be tasty.

Spruce Salt

Makes 1 cup

This salt keeps its color and woodsy aroma indefinitely. Use it on potatoes, smoked fish, game meats, and even beets, whose earthy flavor goes well with spruce.

Note that you can swap the salt in this recipe for sugar and make spruce sugar. The method is the same.

INGREDIENTS

½ cup young spruce tips, brownish sheaths removed

½ cup medium-grain sea salt, kosher salt, or other non-iodized salt

INSTRUCTIONS

1. Coarsely chop the spruce tips. Combine them with the salt, then grind them in batches in an electric coffee grinder (clean your grinder immediately after this use or you may be dealing with gummy spruce residue for months).

2. Freshly made spruce salt is somewhat wet. You can use it as is, and it will store that way just fine. But if you'd like a drier product that's easier to sprinkle, spread the spruce salt out on a baking sheet and dry it in a 150°F oven for 1 hour. Transfer to a food processor and pulse a few times to break up the clumps, or get out your mortar and pestle and regrind the mixture before transferring to an airtight jar.

Spruce Tip–Grapefruit Sorbet

Serves 6

Refreshing and unusual, spruce pairs wonderfully with grapefruit. You'll need to plan ahead, as this recipe starts with letting the spruce steep in a sunny spot for 8 hours. The texture of the sorbet is best if eaten within a few hours of making it. Translation: start the spruce infusion in the morning and serve the sorbet that evening.

INGREDIENTS

2 cups spruce tips, washed and any brown sheaths removed

3 cups water at room temperature

⅓ cup honey

1¼ cups grapefruit juice

INSTRUCTIONS

1. Put the spruce tips into a large, clean, heatproof jar. Pour the water over the spruce and cover the jar. Set in a sunny, warm spot for 6 to 8 hours.

2. Warm the honey and grapefruit juice over low heat in a medium-size pot, stirring to dissolve the honey.

3. Strain the spruce infusion into the pot and stir to combine the ingredients. Remove from the heat, cover the pot, and refrigerate for 1 hour.

4. Transfer the mixture to an ice cream maker and follow the manufacturer's instructions.

Sumac
RHUS TYPHINA (SYN. *R. HIRTA*),
R. CORIARIA, R. AROMATICA, R. GLABRA,
AND *R. COPALLINUM*

Sumac is a good example of how common plant names can be confusing and why we need the scientific names to be certain we're talking about the right plant. Edible sumac species obviously do not include the notorious poison sumac. Instead, edible sumacs provide a lemony liquid extract that can be used to make drinks and sauces; edible shoots; as well as a dried spice that has been in constant use for thousands of years.

Find
Look for sumac shrubs in sunny spots at the edges of fields and roads and on rocky hillsides.

Identify
All of the edible sumacs mentioned above are shrubs that grow anywhere from 3 to 25 feet tall and have alternate, compound leaves made up of pointy leaflets with toothed margins. Winged sumac (*Rhus copallinum*) is the exception, for while it shares all of the other characteristics, its leaflets are not toothed. The leaves of all sumacs turn a brilliant red in autumn.

The small, yellow-green flowers become fuzzy drupes that are each about ⅛ inch in diameter. They are grouped together in upright, cone-shaped clusters that are wider on the bottom than the top. Eventually the drupes turn a rusty red color.

The growth habit and ripe color of the clusters is key to distinguishing the edible sumacs from the notorious poison sumac (*Toxicodendron vernix*), which can give you a vicious rash. Just remember that poison sumac's fruit clusters hang downward and are whitish in color. Edible sumac fruit clusters are upright and rusty-red, although winged sumac's edible red clusters frequently get top heavy and flop over to the side. It's quite easy to tell the poisonous from the safe by remembering these two characteristics.

Rhus typhina, the staghorn sumac, has fuzzy young twigs that lose the fuzz as they mature. Eventually the bark becomes gray-brown with horizontal marks (lenticels) that look like scabs.

R. coriaria, the Mediterranean species sometimes called Sicilian sumac, has smaller leaves and drupes than the other edible sumacs mentioned here. In addition to its use as a spice, it has been used medicinally, as a dye, and to tan leather.

R. glabra, called smooth sumac, looks very similar to staghorn sumac except that its twigs are never fuzzy. It rarely grows taller than 10 feet.

R. copallinum, the winged sumac, has branches that tend to droop toward the earth.

Sustainably Harvest

Harvest sumac's fuzzy fruit clusters once they are a deep, rusty-red color, from summer through fall. When they're at their most sour, the individual berries may have an almost oily sheen to them. Always field-test sumac by picking one of the clusters and licking it—the taste should be strongly sour.

Never harvest sumac immediately after a rain because the water will have washed off the acid that creates the good flavor. A short rain may not have much effect, but all-day rains render sumac tasteless. In arid regions it is sometimes possible to find sumac worth harvesting even in winter after the shrubs have lost their leaves. But this is unlikely in areas that get heavy, regular rainfall or snow.

I like to use pruners to snip off entire clusters, but a pocketknife works well, too. If you are planning to dry the sumac before using it, put the clusters into paper, cloth, or mesh bags. You are not harming the plants by harvesting the fruit clusters.

The young shoots of *Rhus typhina* and *R. glabra* are edible and quite good as long as they are green all the way through with no white pith. Other edible sumacs may have edible shoots as well, but these are the two I have experience with. Break off the new growth from root suckers, tips of branches, or regrowth on sumacs that were cut down. Novice foragers should avoid harvesting and eating sumac shoots until they've seen the fruit clusters and confirmed the shrub they've found isn't poison sumac.

Eat

Dried, powdered sumac berries make a wonderful spice with a deep red color and pleasantly sour taste. Making your own is labor intensive because you have to remove the seeds. Okay, you don't *have* to—commercially sold sumac often includes the ground-up seeds—but I think it's worth it. The seeds and seed shells contribute nothing to the flavor of the final product.

Sometimes sold as "red za'atar," dried sumac seasoning is a spice found throughout the Middle East and in some parts of Europe and northern Africa. It is an essential ingredient in the blend of seasonings called za'atar, which is available everywhere in eastern Mediterranean regions. Instructions for how to make your own dried sumac spice are in the za'atar recipe below.

Interestingly, although people are very familiar with sumac as a spice in the Middle East, no one I have talked to there (including professional foragers) was aware of its North American use as a lemonade-like beverage (and vice versa). With a beautiful reddish color and a refreshingly sour flavor, sumac-ade rivals lemonade anytime in my opinion.

Although it needs some sweetening to turn it into a great beverage, the unsweetened liquid extract of sumac is also useful. It can replace lemon juice in recipes and can be frozen for year-round use.

Sumac-ade

Makes ½ gallon

It seems all the foragers I know have their own version of how to make liquid sumac extract, and of course they swear their method is the best. I've tried every method I've come across, and they all work. I can't claim this method is the "best," but it gives me consistently excellent results.

Iced sumac-ade on a hot summer afternoon . . . you should see the smile on my face just thinking about it!

INGREDIENTS

2 quarts (by volume) whole sumac fruit clusters
2 quarts room temperature or lukewarm water
Honey or sugar syrup to taste

INSTRUCTIONS

1. Put the sumac into a large pot or mixing bowl. Pour the water over and stir. The sumac clusters will float—don't worry about that. Let them soak for 30 minutes, give them a vigorous stirring, then let them soak 30 minutes more.

2. Get in there with your clean hands and rub the drupes, breaking up the clusters as much as possible. Do this for about 2 minutes. Let the whole mess soak for another 30 minutes. Repeat the rubbing and soaking one more time.

3. You should now have a deep red, very sour liquid. Strain it through several layers of cheesecloth, muslin, or a jelly bag. This ensures that you remove not only the obvious solids, but also the fine sumac hairs.

4. Stir in the honey or sugar syrup to taste.

VARIATIONS: SUMAC-ADE POPSICLES OR GRANITA

- Frozen sumac-ade is a wonderful summertime treat. Simply pour the sweetened sumac extract into popsicle molds.

- To make a sumac granita, pour the sumac-ade into a shallow dish such as a pie pan or baking dish. Put it into the freezer and stir with a fork every 25 minutes, pulling the crystals that will form on the sides toward the center. After 1½ to 2 hours, serve or transfer to a freezer container, cover, and return to the freezer.

Za'atar Spice Blend

Makes 1 cup

This savory seasoning can be simply stirred into good extra-virgin olive oil for an extraordinary bread dipping sauce, sprinkled over cooked vegetables, added to salad dressings, or used as a rub for chicken or meat.

What is a bit confusing is that there is a wild Middle Eastern herb called za'atar (*Majorana syriaca*) that is used in the spice blend of the same name. It is related to oregano and marjoram and has a similar, although more intense, scent and flavor. If you can't get the actual za'atar herb to use in this blend, go ahead and make it with oregano or marjoram. But make sure those substitute herbs are garden grown or a high-quality brand: Supermarket-brand herbs won't cut it here. You could also use wild or cultivated *Monarda* leaves.

INGREDIENTS

½ cup dried, powdered za'atar herb (*Majorana syriaca*); dried, powdered oregano or marjoram; or *Monarda* leaves

¼ cup dried, powdered sumac

¼ cup hulled sesame seeds

1 teaspoon medium-grain sea salt

INSTRUCTIONS

1. Combine the za'atar or other powdered herbs with the sumac. If the herb leaves weren't truly powdered but simply crumbled, put them into an electric grinder for 30 seconds. They should be a fluffy powder, not simply crushed leaves.

2. Toast the sesame seeds in a dry skillet over medium heat for 1 minute, stirring constantly. Take care that they don't burn.

3. Combine the toasted sesame seeds and sea salt with the other ingredients. Transfer to a glass jar, cover tightly, and store away from direct light or heat for up to 1 year.

HOW TO MAKE POWDERED SUMAC

1. First dry whole sumac clusters in paper or cloth bags, or lay them out on screens or newspaper or in open cardboard boxes. You want good air circulation around them, so don't pile them up too thickly.

2. After 1 week (or whenever after that you get around to it), strip the drupes off the stems into a food processor. Discard the stems and any leaves.

3. Process for 1 to 2 minutes. This dislodges the sour outer layer (the part you want) from the seeds, but the seed shells are hard enough that they don't break down too much.

4. Transfer the processed sumac to a fine mesh sieve over a bowl. Rub the sumac through. The crumbly red spice should come through the sieve, leaving the more solid seed fragments behind. It's okay if a bit of crushed seed comes through. You can use your sumac spice as is, or regrind it into a finer powder in an electric coffee grinder.

Za'atar Spice Blend

Violet
VIOLA SPECIES

Shade-loving violets are lovely to look at in early spring, when their edible purple flowers catch our eyes. But the leaves are also good to eat and have a longer harvest season than the flowers. Use them in spring and summer as salad, soup thickener, and a cough remedy.

Find

Violets can tolerate full sun, but are usually found growing in partial shade. They prefer moist soil and thrive under deciduous trees, where they make the most of the early spring sunlight coming through the still winter-bare branches. Later in the year, the plants get relief from hot summer sunshine when the trees they are growing under leaf out.

Identify

Learn to identify violet leaves so that you can recognize the plants even when they are not in flower. The heart-shaped leaves grow in a rosette. They have fine teeth along the margins and pointed tips. Because garlic mustard likes similar shady, disturbed soil situations and also has a basal rosette of heart-shaped leaves, novice foragers sometimes confuse them when the flowers aren't present. But a violet leaf's tip is sharply pointed, as are the tiny teeth on the margins, whereas garlic mustard's basal leaves have scalloped, rounder edges. The shades of green are different, too, as are the venation patterns (turn a violet leaf over and you'll clearly see the prominent

veins, especially in bigger leaves). But really all you have to do is use your nose: garlic mustard leaves smell like garlic and mustard, whereas violet leaves don't really have a smell.

Young violet leaves are curled in on themselves like scrolls rolled in from both sides. This is the ideal stage to harvest them.

The first flowers that wild violets produce in early spring are the showy ones that are so pretty on salads. They are usually purple with some white near the center, but sometimes they are mostly white. These sterile flowers are about ¾ inch in diameter and grow on narrow, leafless stalks that can be several inches long. There are five petals on these flowers, and the side petals have white hairs at their bases.

In summer, violets produce self-pollinating, petal-less flowers that you probably won't notice. These become three-parted capsules that eject the small, round seeds.

Violet roots are knobby, branching, somewhat horizontal rhizomes. They are not edible.

Sustainably Harvest

Violet's sterile purple or purple and white flowers can be collected at any time during the early spring flowering season. Pinch them off with their long, thin flower stalks attached, as these are also tasty.

The leaves are good spring through summer. As the plants mature, it is best to harvest only the smaller, partially furled leaves, as the bigger leaves can get tough and stringy.

Violets are perennial plants that will regenerate from the inedible root that you leave in the ground.

Eat

Violet's pretty first flowers (the ones that don't produce seeds) are delightful in salads, as are their stems. (Thanks to my mom for pointing out that the flower stems have their own taste that is distinct from that of the flowers and leaves.) The young leaves are also excellent in salads.

The purple flowers turn syrup and sugar pale indigo; it's a natural food coloring that makes elegant drinks and desserts.

Cooked, violet leaves reveal their mucilaginous quality, which isn't very noticeable in the raw leaves. This means a heap of them cooked on their own will be a slimy mass. But it also means that they can be used to thicken soups and stews. For this purpose, I use them either fresh and finely chopped, or dried and powdered.

That same mucilaginous property makes violet leaves a traditional herbal remedy for the respiratory system. Dried and made into a mild-tasting infusion, violet leaves are soothing for coughs, especially dry, ticklish coughs.

Violet Leaf, Flower, and Stem Salad

Serves 4 as a side dish

Make this salad in early spring during the few weeks when you can harvest the purple flowers, their stems, and the most tender leaves. The dressing is deliberately mild to showcase the gentle flavors and different textures of the various violet parts.

INGREDIENTS

1 quart young violet leaves, flower stems, and flowers

Optional toppings: pine nuts or walnuts, mild goat cheese

2 tablespoons sunflower or other neutral-tasting vegetable oil

1 teaspoon balsamic vinegar

1 teaspoon champagne or sherry vinegar

½ teaspoon light honey such as clover or wildflower

Salt to taste

INSTRUCTIONS

1. Rinse the violet parts and dry them in a salad spinner or by rolling them up in a clean dish towel. Separate the flowers from their stems.

2. Divide the leaves between four salad plates, then top with the flowers and flower stems. Sprinkle on the optional toppings, if using.

3. Whisk together the oil, vinegars, honey, and salt. Drizzle the dressing over the salads and serve immediately.

Violet Sugar

Makes 1 cup

This treat is all about the striking naturally purple hue the sugar takes on from the violets. Use violet sugar to decorate cookies, cakes, and shortbread, or in lemonade and iced tea.

INGREDIENTS

⅔ cup lightly packed violet flowers

1 cup sugar

INSTRUCTIONS

1. Pinch the calyxes (green parts at the base of the flowers) off the violet flowers.

2. Rinse the petals in a sieve, then spread them on a paper or cloth towel and pat mostly dry.

3. Put the sugar into a bowl or mortar and spread the still slightly damp violet flowers over the top of the sugar. Rub the violets into the sugar with a pestle, a clean rock, or the bottom of a sturdy bottle or glass. Use a fork to break up the petals even more. The sugar will start to turn purple as you do this, which is what you want.

4. Cover the bowl or mortar with a dry dish towel and leave the violet sugar to dry out for a few hours or as long as a day. Transfer to an airtight container.

Wild Garlic

ALLIUM VINEALE, A. CANADENSE, A. NEAPOLITANUM, AND *A. URSINUM*

Wild *Allium* species grow throughout the Northern Hemisphere as well as in some parts of South America and Africa. It's hard to imagine most savory dishes without some sort of allium (onions and cultivated garlic are also in this genus). The wild garlics are just as essential in a forager's kitchen.

Find

Look for wild garlics at woodland edges and growing under deciduous trees. In winter and early spring they may be growing in full sun because the branches over them are bare of leaves. But by midspring, once the trees leaf out, wild garlics are usually growing in dappled light or partial shade.

Identify

Let's start with the most important identification characteristic of all edible wild garlics: They smell like garlic! I'll get into the details of leaf shape, etc., in a moment so that you know what you are looking for. But once you think you've spotted a wild garlic, always confirm your ID by crushing part of the plant. If it smells like garlic, it's safe to eat.

All alliums have flowers that look like globular pom-poms of small florets. Sometimes the plants skip flowering altogether, and instead you'll see clumps of green or reddish bulblets growing on the tips of stalks. These stalks are noticeably stiffer than

the leaves. The bulblets frequently have tiny, stringy leaves growing out of them.

Allium vineale, commonly known as field garlic, is an invasive weed with narrow, hollow leaves that lead many to think they've found some sort of wild chives. Also called onion grass, it grows in clumps and is extremely cold hardy, making it a good winter forage throughout its range in eastern North

America. The small, intensely garlic-scented underground bulbs are usually the size of a pinky fingernail, occasionally larger.

A. canadense, also called meadow garlic or Canadian garlic, has a milder scent and tastes more like an onion than garlic. It also grows in eastern North America, but can be found in the Midwest and down into Cuba and Central America as well. It has narrow, grass-like leaves.

A. neopolitanum, or white garlic, has flat, strap-like leaves and grows in the Middle East, California, Florida, Georgia, Alabama, and all around the Mediterranean. It has a gentle flavor reminiscent of eastern North America's ramps (*A. tricoccum*).

A. ursinum is also very similar to *A. neopolitanum* and *A. tricoccum* in both its appearance and its mild, leek-like flavor. Also called ramsons and bear garlic, it has flat, lily of the valley–like leaves and a broad range that includes Asia and Europe. Because its "look-alike" lily of the valley is toxic, it's worth pointing out the differences: Bear garlic leaves emerge singly from the base of the plant, whereas lily of the valley leaves emerge as two or three leaves on the same stem. More importantly and unmistakably, lily of the valley does not smell at all of onions or garlic, but *A. ursinum* does.

Sustainably Harvest

Some wild garlics are invasive in their current range, whereas others may be in danger of overharvesting. Only harvest where the plants are abundant, and if you aren't sure whether the species you've got is invasive or endangered, only harvest the aboveground parts. Only if you know you've got an invasive allium like field garlic should you go ahead and dig up the underground bulbs, too.

Alliums are ephemerals, which means they are visible aboveground only for certain months of the year. Look for wild garlic in August and you will be disappointed. The plants put out leaves during the cool temperatures of autumn and can sometimes be harvested straight through the winter. They put out a vigorous burst of growth in early spring and then produce flowers or bulbils. Most of this happens while the branches of the deciduous trees they like to grow under are still bare. By the time the weather is getting hot and the trees have leafed out, *Allium* species have died back to the ground.

Eat

Wild garlics have edible leaves, bulbs, and flowers or bulbils. They are all delicious raw or cooked in any way that you would use cultivated garlic or chives.

The flowers can be infused in vinegar to make an excellent salad dressing ingredient. If the leaves get too tough to chew, simply use what I call the bay leaf method to extract their flavor: tie a bunch of them in a knot, drop the garlic leaf knot into any broth or soup, then discard it before serving.

Skordalia

Makes about 3 cups

Skordalia is a Greek spread that can be made with potatoes, breadcrumbs, or both. It is intensely garlicky . . . in a good way (*skorda* means "garlic" in Greek). Depending on which wild garlic you use, it may be very pungent (*Allium vineale*) or fairly gentle (*A. neopolitanum*).

INGREDIENTS

1 pound potatoes (about 3 medium potatoes), peeled

½ cup potato cooking water

1 cup fresh breadcrumbs (traditionally this was a recipe for using up stale bread)

3 tablespoons minced wild garlic bulbs or leaf bases

2 tablespoons minced wild garlic leaves

½ cup extra-virgin olive oil

¼ cup lemon juice

Salt and freshly ground black pepper

INSTRUCTIONS

1. Chop the peeled potatoes into chunks and put them into a pot. Cover them with cold water, then bring to a boil over high heat. Reduce the heat and simmer until the potatoes are fall-apart tender when pierced with a fork.

2. Reserve ½ cup of the cooking liquid. Drain the rest through a colander. Return the potatoes to the pot along with the breadcrumbs, garlic bulbs and leaves, olive oil, and lemon juice. Use a potato masher, the bottom of a wine bottle, or a large fork to mash the ingredients together. You can keep mashing until the mixture is fairly smooth or leave it with a rougher texture. Whatever you do, do not be tempted to use a food processor or your skordalia will have the consistency of glue. Add the reserved cooking water a little at a time until the skordalia reaches a spreadable thickness. Mix in salt and pepper to taste.

3. Wait at least 1 hour before serving for the garlic flavor to permeate the other ingredients. Serve at room temperature with crackers or crudités.

Wild Garlic Powder

Makes about ¼ cup if using 1 cup field garlic

You'll get the best results with this if you have a dehydrator, but it's totally possible to make garlic powder in an oven as well.

INGREDIENTS

Field garlic or other wild garlic species

INSTRUCTIONS

1. Clean, peel, and mince the bulbs of field garlic (*Allium vineale*) or other wild garlic species. Honestly, this is the most tedious part of any garlic recipe, wild or cultivated. I highly recommend a good garlic press that you can put the small bulbs into, peels and all. The garlic will get pressed through, leaving the skins behind, which means you can skip the peeling. However, only a sturdy garlic press with good leverage will do this—a flimsy plastic model will be worthless.

2. Dehydrator method: Line a tray or trays of your dehydrator with something to prevent the garlic from falling through the tray's holes. This could be a liner that came with your dehydrator, the silicone sheets that can be ordered for various dehydrator models, or parchment paper that you cut to size.

    Spread the minced or pressed garlic on the sheet in a thin layer. Dry at 120°F until crispy-dry but not burnt. This can take several hours; overnight is okay. (You are dehydrating the garlic, not cooking it, so it won't burn.) Warning: your home will smell garlicky.

3. Oven method: Line a baking sheet with parchment paper. Spread the minced or pressed garlic in a thin layer on the sheet and bake at 150°F or the warm setting of your oven. Use a long-handled wooden spoon or a spatula to turn the garlic after 15 minutes. Continue baking until crispy-dry but not burnt. This takes much less time on even the lowest oven setting than in a

dehydrator, so check every 10 minutes or so to see if it is done.

4. Whether using a dehydrator or the oven, the next steps are the same. Take the trays or sheets of garlic out and let cool at room temperature for 5 minutes. Grind the dried garlic in an electric grinder or with a mortar and pestle. Store in an airtight container.

Wild Grape
VITIS SPECIES

From stuffed grape leaves to grape jelly, pickling to winemaking, grapes are one of the most useful wild foods.

Find

Wild grapes clamber over trees, shrubs, and fences in full to partial sun locations. They are common in and on thickets, near streams and riverbanks, and near shores.

Identify

Grapes are woody vines with leaves that can grow as large as 8 inches across. The maple-shaped leaves have three lobes with undersides that are usually felt-like. The plants have forked tendrils that hook onto other plants and structures to climb toward more sunlight. The clusters of small, green-ish flowers are aromatic, and each flower has five petals. These flowers turn into the familiar-looking clusters of round fruit. Wild grapes are usually deep purple when ripe. Each grape has several seeds.

Grapes have several poisonous look-alikes, and the shape of the leaves is not enough for a positive ID. You also need to look for the forked tendrils, woodiness of the vine, and multiple seeds in each fruit. One or another of these characteristics is absent in each of the poisonous look-alikes. But if *all* of these characteristics are present, you've found a wild grapevine.

Sustainably Harvest

Snipping off a few leaves or clusters of fruit will not harm this perennial plant. It's best to graze when you're harvesting the leaves, taking just one or two from each section of vine rather than denuding an entire strip.

Harvest the leaves in spring when they are at least as big as the palm of your hand but before the plants set fruit. After that the veins become too tough and chewy.

Harvest the fruit when it is full size and ripe (unless you are harvesting the unripe grapes to make verjus). If you're not sure if the grapes are ripe yet, simply pop one in your mouth and let your taste buds guide you.

Both individual leaves and fruit clusters can be snapped off by hand, or you can use pruners.

Eat

I don't think I need to tell you how to eat a juicy, freshly picked, raw grape (although with wild ones you will need to spit out the seeds). So what else can you do with them?

Wine, of course, and I've included some home winemaking resources in the Useful Resources section. Grape jelly is a classic. And unfermented grape juice is good and can be preserved through canning or freezing. However, for any of these uses you first need to deal with the high level of tartrate in the fresh juice, which can produce an unfortunate burning sensation. To get rid of the tartrate, simply let the juice sit for a couple of days. Pour off the juice, leaving behind the layer of tartrate gook that will have settled to the bottom of the container. Doing this twice does a better job of separating out the tartrates than just once.

Grape leaves are fantastic for making dolmades (stuffed grape leaves), and you'll find that the wild leaves contribute much more flavor than those you can buy in jars at the store. In fact, they make an interesting seasoning in their own right, as in the mushroom recipe below. Grape leaves also help homemade pickles keep their crunch: Just add one to each jar of pickles. You also can tuck one in over the top of a jar of lacto-fermenting food to keep the vegetables submerged in the brine.

The "bloom" or whitish coating on grapes is wild yeast. This is one of the factors that made grapes *the* fruit traditionally used for wine—they didn't need added yeast for fermentation to occur. You also can use those wild yeasts on grapes to make sourdough starter.

Last but not least, woody trimmings from grapevines make excellent chips for smoking foods in a smoker or over the coals of a campfire.

Mushrooms with Grape Leaves

Serves 4

I first came across this recipe in Elizabeth David's *Italian Food,* and apparently she got it from Edmond Richardin's *L'Art du Bien Manger* (1913). It probably goes back much further than this. The special delight of this recipe is that it transforms ordinary supermarket button mushrooms into something as earthy and enticing as some of the best wild mushrooms. But if you really want an over-the-top fabulous dish, make this with wild oyster or chicken of the woods mushrooms.

INGREDIENTS

1 quart small, whole button mushrooms or thickly sliced, larger mushrooms

Enough fresh grape leaves to cover the bottom of a baking dish in a single layer

1 to 1½ cups extra-virgin olive oil

1½ teaspoons coarse sea salt

8 whole, unpeeled garlic cloves

½ teaspoon ground black pepper

INSTRUCTIONS

1. Preheat the oven to 325° F.

2. Rinse the mushrooms. Dry by spinning them in a salad spinner or by rolling them up in a clean dish towel.

3. If using button mushrooms, slice the stems away and chop the stems. Set the chopped stems aside. Leave small mushroom caps whole; cut large caps into halves or quarters.

4. Line a baking dish with the grape leaves. Pour in enough of the olive oil to completely cover the leaves.

5. Place the baking dish in the oven and let the grape leaves cook until their color changes from bright green to a drab khaki color—about 5 minutes.

6. Arrange the mushrooms over the grape leaves. If using button mushrooms, place them stem-side up. Sprinkle the salt over the mushrooms.

7. Bake for 30 minutes.

8. Scatter the reserved, chopped mushroom stems over everything and press them into the partially cooked mushrooms with the back of a spoon. Push the garlic cloves in between the mushrooms.

9. Return the dish to the oven and bake for another 10 minutes. Season with freshly ground black pepper.

10. If you made this dish with the tender young grape leaves of spring, serve them along with the mushrooms because they are delicious, too. Older grape leaves may be too tough to eat, but will still impart their unique taste to the mushrooms. Spoon some of the oil that the leaves and mushrooms cooked in over each serving (any remaining oil and juices make a great sauce for another dish). Serve hot or at room temperature.

Wild Grape Sourdough Starter

Makes 1 cup

This starter is infallibly lively.

You can use it to add amazing sourdough flavor to breads and pancakes, but its real purpose is to replace packaged yeast. Maintaining a sourdough starter does take a certain amount of responsibility because it needs to be "fed" regularly, but the results are absolutely worth it. Plus there's something delightfully self-sufficient about making your own yeast rather than buying it.

INGREDIENTS

1 pound unwashed, ripe wild grapes
1 cup flour (whole-wheat or all-purpose)
1 cup bread flour

INSTRUCTIONS

1. Remove the grapes from their stems and place them in a bowl. Crush them well with a potato masher or the bottom of a wine bottle. Cover with cheesecloth or a clean dish towel and leave at room temperature for 3 days, stirring vigorously at least twice a day.

2. The starter will start to show the frothy, bubbly signs of a successful fermentation thanks to the wild yeasts on the grapes. Strain the liquid, discarding the skins and seeds. Return the liquid to the bowl and stir in 1 cup of flour. Cover with cloth as before and leave at room temperature for 24 hours.

3. Measure the starter and save just 1 cup of it. Add the bread flour (you can get away with regular all-purpose flour at this stage if you don't have bread flour). The mixture should have the consistency of a thick batter, but not as thick as dough.

Feel free to add more flour if it is too thin, or some water if it seems too thick. Repeat this step the next day, and again the one after that. Each time you will need to discard some of the starter so that you begin by working with just 1 cup of it (you can compost the discards).

4. Transfer the starter to a loosely covered jar in the refrigerator. Feed it once a week by adding 1 cup flour (ideally bread flour) and 1 cup water to 1 cup of your active wild grape sourdough starter. Meanwhile, use that starter instead of or in addition to baking yeast for amazing bread, muffins, waffles, and pancakes. Check the Resources section for websites with excellent instructions and recipes using sourdough starter.

Mosto Cotto (Grape Molasses)

Makes 1 to 2 cups

Also called *vino cotto, sapa*, and *petimezi, mosto cotto* means "cooked must." The "must" is smashed grapes that in this case are cooked down into a delectable sweet sauce with the consistency of maple syrup. It is heavenly on fruit, cheese, or in salad dressings.

This is an ancient recipe found in countries all around the Mediterranean. It was originally used to create a sweetener in the era before sugarcane was introduced to that part of the world. Wine grapes were usually used to make *mosto cotto*. They contain as much as 25 percent natural sugars, which is much higher than the sugar content of most wild grapes and even than that of store-bought table grapes. If you like (and if your grapes are tart), you can add ½ to 1 cup of sugar to the wild grape must in this recipe to approximate the sweetness of wine grapes. You can still make the *mosto cotto* without the added sugar, but you will end up with much less of it.

INGREDIENTS

10 pounds ripe wild grapes, removed from their stems (red or purple grapes will give the best color)

1 tablespoon grapevine ash (optional but traditional; see instructions below)

Authentic recipes for *mosto cotto* almost always include some ash left from burning the woody parts of grapevines. You can leave this out and it will still be good, but if you want to make it the traditional way, simply make a small fire out of dried woody grapevines and let it burn until reduced to ash. Sift the ash through a sieve and discard any large particles. You'll only need 1 tablespoon of the ash for this recipe, so you might want to save any surplus for making *mosto cotto* in future.

Tip: Although true balsamic vinegar is aged in oak casks, you can make a good approximation by combining *mosto cotto* with some red wine vinegar. It will be almost as good as the (expensive) real thing, and much, much better than cheap supermarket brands of balsamic.

INSTRUCTIONS

1. Smash the grapes. No, you do not need to stomp them with your feet. (Okay, I did do that once, but that's another story!) A potato masher and some elbow grease will work just fine. Or put them into a food processor in small batches and pulse a few times. (You don't want to purée the grapes, just crush them.)

2. Put the smashed grapes along with the vine ash (if using) into a large pot over medium heat. Let the temperature rise until it is steaming hot but just below a simmer. (It should not be bubbling much if at all.) Maintain this approximate temperature and cook, stirring often, for at least 1 hour. (During this hour the pectin in the skins and seeds of the grapes will permeate the juice.)

3. Strain the must through cheesecloth. Squeeze the bundle of cloth or press on it with the back of a wooden spoon to get out as much liquid as possible. Return the strained liquid to the pot.

4. Continue to cook the *mosto cotto* down until it is the consistency of olive oil or thick maple syrup. This can take a couple of hours. Again, do not let the liquid boil or your final product will have a caramelized "off" taste.

5. Remove from the heat and let cool slightly before bottling. *Mosto cotto* will keep in a cool place for at least 1 year.

Wild Lettuce
LACTUCA SPECIES

Although related to our cultivated lettuces, wild lettuces have their own unique taste. Not just for salads, they are also a good cooked vegetable.

Find
You'll find *Lactuca canadensis* and *L. biennis* growing in partial shade in forests, parks, and along clearing edges.

    L. serriola, prickly lettuce, prefers full sun in disturbed soils, usually near human habitation.

Identify
Wild lettuces start out with a basal rosette of 5- to 10-inch-long leaves that remind many people of dandelion. However, dandelion never sends up a tall flower stalk that has leaves on it, and wild lettuces do. The leaves on the flower stalk are smaller than the basal rosette leaves.

    As is true with dandelion, white sap oozes out of any part of a wild lettuce plant when you break it.

    The yellow ray flowers also resemble dandelion, although dandelion flowers are much bigger.

    Wild lettuces are also often confused with sow thistles (*Sonchus* species). Wild lettuces have prickles or hairs along the underside of their leaf midribs that sow thistles lack.

    Lactuca canadensis is the tastiest of all the wild lettuces. It most often has deeply lobed rosette leaves with smaller, usually unlobed flower leaves on the flower stalks. The latex is tan or brown if exposed

to the air. *L. canadensis* sends up a flower stalk that, although unbranched, is quite different from that of dandelion. *L. canadensis* stalks can grow as tall as 9 feet. Although the flowers look like miniature dandelions, they grow in clusters on the tall stalks rather than a single flower to each stalk.

    L. biennis, or bitter lettuce, has white latex and grows as tall as 11 feet when in flower. Its flowers are white rather than yellow like other wild lettuces.

L. serriola, or prickly lettuce, is a shorter plant even when flowering, with lobed to barely lobed leaves. These leaves have prickles on the edges and underside of the midrib. Prickly lettuce leaves rotate to face the sun. The flowers are yellow.

Sustainably Harvest

These are common weeds and will not be endangered even if you collect quite a lot of them.

Collect the leaves and tender stalks before the plants flower. I think they are at their best just as the flower stalks begin to elongate. In older plants, not only does the taste become too bitter to be enjoyable, but in some species the line of prickles on the midrib becomes so stiff that the texture isn't pleasant. Stick with the young, soft leaves and stalks.

Pick the unopened flower buds as they appear.

Eat

Although salad is the first thing most people think of when they hear "wild lettuce," I use it just as often as a cooked green. If you collect it young enough, it's not too bitter, and the faint bitterness that is there is a desirable part of its flavor profile. It pairs well with sweet-sour flavors like balsamic vinegar and salty-umami ones like bacon or cheese.

The unopened flower buds can be eaten raw or cooked, or pickled.

Five Tastes Salad

Serves 2 as a main course, 4 as a side

This salad includes all the five flavors our taste buds are able to detect. Wild lettuce provides mild bitterness, the dressing brings light sourness, the raisins, sweetness, and the cheese and bacon both salt and umami flavors (leave the bacon out for a vegetarian version). In addition, there are contrasting textures from crunchy to creamy to chewy. The result is a main course salad that I enjoy as an entire meal by itself.

Depending on what's in season, feel free to add other wild vegetables and toppings. Edible flowers such as spicy yellow mustard flowers and pretty pink mallow blooms are good and are often in season when the wild lettuce plants have not yet flowered and are still good to eat. You could also substitute other dried fruits for the raisins—any dried fruit with a chewy texture and sweet flavor will work here.

INGREDIENTS

¼ cup plus 2 tablespoons mayonnaise

2 tablespoons balsamic vinegar

2 tablespoons wine vinegar

1 tablespoon wild mustard seed prepared mustard or Dijon mustard

1 teaspoon honey

½ teaspoon dried thyme

½ teaspoon freshly ground black pepper

Pinch of salt

½ cup sunflower seeds or chopped walnuts (only use ¼ cup if you are using wild black walnuts)

4 ounces bacon (optional), sliced into small pieces

1½ pounds wild lettuce

½ cup raisins

½ cup crumbled blue cheese

INSTRUCTIONS

1. Whisk together the mayonnaise, balsamic and wine vinegars, mustard, honey, thyme, black pepper, and salt.

2. Toast the sunflower seeds or walnuts in a dry skillet over medium heat, stirring or tossing frequently, for about 2 minutes until they are fragrant but not burnt. Take the nuts or seeds out of the skillet and set them aside. In the same skillet, cook the bacon pieces until they are just starting to become crisp. Remove the bacon pieces and let them drain on a paper towel.

3. Depending on the wild lettuce species and the stage at which you harvested, those prickles on the underside of the midrib may be ignorable or not. If they are too bristly, tear the tender sides of the leaves away from the midrib and use only those. Otherwise, just tear the leaves into bite-size pieces.

4. In a large bowl, toss the wild lettuce with the dressing. Divide the dressed greens between two (main course) or four (side dish) plates. Top with the reserved toasted nuts or seeds, bacon, raisins, blue cheese, and any edible wild flower garnishes. Serve immediately.

Wood Sorrel

OXALIS SPECIES

Wood sorrel's leaves, flowers, and seedpods all have such a great sour flavor that the kids who come on my foraging tours call them "wild pickles."

Find

Wood sorrels grow in partial sunlight in moist woodlands and in disturbed garden soil. You might find them growing in full sun, but those plants are usually too small, stringy, and tough to bother with unless they are also getting a lot of irrigation.

Identify

Wood sorrels are often mistaken for clover because of their trefoil leaves. But if you compare the two, you'll see that wood sorrel's leaflets are heart-shaped, completely hairless, and lack the chevron marking that clover leaves often have. They also have a pleat down the center and close up like books at night and sometimes during the heat of midday (clovers don't do this). Some species have speckles on their leaves (*Oxalis pes-caprae*).

Once the plants are in flower, there's no confusing *Oxalis* species with pom-pom-flowered clo-

ver. Five-petaled wood sorrel flowers can be yellow (*Oxalis stricta, O. pes-caprae*) or pink (*O. montana*), depending on the species.

The seedpods look like miniature okra (I call them "fairy okra"). They grow on thin stalks at quirky angles and are slightly hairy.

Some wood sorrel plants spread by rhizomes, so if you pull up what appears to be one plant you'll often find its neighbor coming up as well.

Sustainably Harvest

Wood sorrels are often prolific to the point of being considered invasive in some terrains. You do not have to worry about overharvesting, and in any case the plants will regenerate if you gather only their upper aerial parts.

My favorite way to harvest wood sorrel is to gather a bunch in one hand while I snip the stems about halfway down the plant with scissors or prun-ers. Still grasping the bottom end of my wood sorrel bouquet with one hand, I hold it over a bowl or container. With my other hand, I encircle the bunch and gently pull from base to tip. The tender leaves, pods, and flowers come away, leaving the unpalatably wiry stems behind.

Eat

Wood sorrel's tangy taste is excellent raw or cooked. Raw, wood sorrel is a terrific addition to salads and spreads. Cooked, it makes a great sauce for vegetables, whole grains, or seafood. Note that when cooked, the leaves change from emerald green to a drab olive green. There is no way to prevent that, but fortunately it doesn't affect the flavor.

Wood sorrel can be used in any recipe that calls for garden or French sorrel or sheep sorrel. Even though it isn't related to those plants, its flavor profile is identical.

Wood Sorrel Soup

Serves 2 as a main dish, 4 as an appetizer

This is an easy soup recipe that is as terrific served cold as hot.

INGREDIENTS

1 tablespoon butter or extra-virgin olive oil

1 pint lightly packed wood sorrel leaves, green pods, flowers, and stems (tender ones only)

1 teaspoon wild or cultivated garlic, minced

1 quart chicken or vegetable stock

½ pound potatoes, peeled and chopped into chunks

Salt and freshly ground black pepper

INSTRUCTIONS

1. Heat the butter or olive oil in a pot over medium heat. Add the sorrel (save a few leaves and flowers for garnish). Cook, stirring, until all the leaves have wilted and the color has changed from bright to drab green.

2. Add the garlic and cook, stirring constantly, for 30 seconds.

3. Add the stock and the potatoes. Bring to a boil over high heat. Reduce the heat and simmer, partially covered, until the potatoes are soft enough to fall apart when pierced with a fork.

4. Remove from the heat and let cool for 5 minutes. Purée right in the pot with an immersion blender or in two batches in a blender. Add a little more stock if it seems too thick. Add salt and freshly ground pepper to taste. Serve hot or cold, garnished with wood sorrel leaves and flowers.

Other plants that are good in this recipe include sheep sorrel and Japanese knotweed. You can also substitute peppergrass or shepherd's purse seeds for the black pepper.

PLANT, MUSHROOM, AND SEAWEED IDENTIFICATION SITES

identifythatplant.com

mushroomexpert.com

plants.usda.gov

rogersmushrooms.com

seaweed.ie

WILD EDIBLES WEBSITES AND BLOGS

downanddirtygardening.com

eattheweeds.com

ediblewildfood.com

fat-of-the-land.blogspot.com

firstways.com

foragersharvest.com

honest-food.net

hungerandthirstforlife.blogspot.com

ledameredith.com

motherearthnews.com (not exclusively a wild edibles site, but the author writes a foraging blog for them, and there are also many other writer's wild plant and mushroom articles on the site)

returntonature.us

sergeiboutenko.com

wildfoodadventures.com

wildfoodgirl.com

wildmanstevebrill.com

ONLINE EDIBLE WILD PLANTS GROUPS

On Facebook

Edible Wild Plants; Edible Wild Plants/Mushrooms And The ID Of Plants, Trees And Mushrooms; Foragers Unite!; Foraging Wild Edible Plants and Medicinal Herbs; Hunt Gather Cook; Merriwether's Foraging Texas; Midwest Wild Edibles & Foragers Society; Return to Nature; Wild Edible Foods; Wild Edibles; Wild Edibles Of The Finger Lakes; Wild Edibles & Healing Herbs with Heather Gardner; Wild Edibles of Missouri; Wild Edibles with Sergei Boutenko; Will Forage for Food

On Yahoo! Groups

ForageAhead, PlantForagers

FORAGING BOOKS AND FORAGING APPS

Bennett, Chris. *Southeast Foraging: 120 Wild and Flavorful Edibles from Angelica to Wild Plums.* Portland, OR: Timber Press, 2015.

Blair, Katrina. *The Wild Wisdom of Weeds: 13 Essential Plants for Human Survival.* Vermont: Chelsea Green Publishing, 2014.

Boutenko, Sergei. *Wild Edibles: A Practical Guide to Foraging, With Easy Identification of 60 Edible*

Plants and 67 Recipes. Berkeley, CA: North Atlantic Books, 2013, and Sergei's Wild Edibles app, both available via his website, https://sergeiboutenko.com.

Brill, "Wildman" Steve. *Identifying and Harvesting Edible and Medicinal Plants in Wild (and Not So Wild) Places.* New York: William Morrow, 1994, and Steve's Wild Edibles app, both available via his website, wildmanstevebrill.com.

Deur, Douglas. *Pacific Northwest Foraging: 120 Wild and Flavorful Edibles from Alaska Blueberries to Wild Hazelnuts.* Portland, OR: Timber Press, 2014.

Thomas Elias and Peter Dykeman. *Edible Wild Plants: A North American Field Guide to Over 200 Natural Foods.* New York: Sterling, 1982.

Thomas Elpel and Kris Reed. *Foraging the Mountain West: Gourmet Edible Plants, Mushrooms, and Meat.* Pony, MT: HOPS Press, 2014.

Dina Falconi and Wendy Hollender. *Foraging & Feasting: A Field Guide and Wild Food Cookbook.* Accord, NY: Botanical Arts Press, 2013.

Merritt Fernald, Alfred Kinsey, and Steve Chadde. *The New Edible Wild Plants of Eastern North America: A Field Guide to Edible (and Poisonous) Flowering Plants, Ferns, Mushrooms and Lichens.* Self-published by Steve Chadde, 2014. Email: steve@chadde.net.

Kallas, Dr. John. *Edible Wild Plants: Wild Foods from Dirt to Plate,* Layton, UT: Gibbs Smith, 2010. Order a print copy via wildfoodadventures.com or the Kindle edition via amazon.com.

Lincoff, Gary. *The Joy of Foraging: Gary Lincoff's Guide to Finding, Harvesting, and Enjoying a World of Wild Food,* Quarry Books, 2012; and *The Complete Mushroom Hunter: An Illustrated Guide to Finding, Harvesting, and Enjoying Wild Mushrooms.* New York: Crestline Books, 2011.

Meredith, Leda. *Northeast Foraging: 120 Wild and Flavorful Edibles from Beach Plums to Wineberries.* Portland, OR: Timber Press, 2014.

Peterson, Lee. *A Field Guide to Edible Wild Plants of Eastern and Central North America.* New York: Houghton Mifflin, 1977.

Rose, Lisa M. *Midwest Foraging: 115 Wild and Flavorful Edibles from Burdock to Wild Peach.* Portland, OR: Timber Press, 2015.

Rothkranz, Markus. *Free Food and Medicine: Worldwide Edible Plant Guide, First Edition.* Self-published, 2012. www.FreeFoodandMedicine.com.

Thayer, Samuel. *The Forager's Harvest: A Guide to Identifying, Harvesting, and Preparing Edible Wild Plants,* Forager's Harvest Press, 2006; and *Nature's Garden: A Guide to Identifying, Harvesting, and Preparing Edible Wild Plants.* Bruce, WI: Forager's Harvest Press, 2010.

Zachos, Ellen. *Backyard Foraging: 65 Familiar Plants You Didn't Know You Could Eat.* North Adams, MA: Storey Publishing, 2013.

TOOLS AND USEFUL INFORMATION

The DaveBilt Nutcracker is great for cracking acorns and other thin-shelled nuts, especially if you are processing them in quantity: shop.davebilt.com/ Davebilt-43-Nutcracker-43.htm.

Grapestompers has great information on home winemaking and home brewing, as well as every kind of equipment you might need for those pursuits: http://grapestompers.com.

Sourdough Home has good recipes and information on using sourdough starter (once you've made yours from wild grapes): sourdoughhome.com.

Tap My Trees has detailed information on how to tap trees for their sweet sap. They also sell spiles and other supplies for tapping maple, birch, and other trees: tapmytrees.com.

For information on boiling water bath canning, the gel point in jelly-making, high altitude canning, and other food preservation methods, head over to foodpreservation.about.com, or pick up a copy of Leda Meredith's *Preserving Everything: How to Can, Culture, Pickle, Freeze, Ferment, Dehydrate, Salt, Smoke, and Store Fruits, Vegetables, Meat, Milk, and More*, Countryman Press, 2014.

ACKNOWLEDGMENTS

While I was writing this book I traveled on three continents and stayed with many gracious and adventurous hosts. They let me forage in their backyards, forests, gardens, and neighborhood parks. They also let me test recipes in their kitchens and (I hope) enjoyed tasting the results. Huge thanks to Dimitra Chalkiadaki, Penelope and Frank Coberly, Marc der Kinderen, Paige Irvine, Kelly Johnson, Debbie Marcum Midkiff, Laura and Ken Orabone, Maya and Roey Orbach, Francis Patrelle, Melissa Price, Pascal and Buddy Rekoert-Valdez, Roger Rynlds, Sam Thayer, Raquel and Tomislav Valiente de Kostadinov, Petra van Noort, Ellen Zachos, and anyone else who hosted me!

Gratitude and respect to Mike Krebill, who not only brought his keen editorial eye to this manuscript, but also his foraging expertise.

Thanks also to Liz Neves for the hawthorn and the peppergrass, and Jeremy Umansky for the mushroom pics.

Biggest thanks of all to Ricky Orbach, my husband, foraging companion, and most honest and enthusiastic taste-tester. Here's to sharing many more years of wild harvests.

A lifelong forager (it's her great-grandmother's fault), Leda Meredith is the author of four previous books, including *Northeast Foraging: 120 Wild and Flavorful Edibles from Beach Plums to Wineberries* and *Preserving Everything: Can, Culture, Pickle, Freeze, Ferment, Dehydrate, Salt, Smoke, and Store Fruits, Vegetables, Meat, Milk, and More.* She is an instructor at the New York Botanical Garden and at the Brooklyn Botanic Garden, specializing in edible and medicinal plants. You can follow Leda's sustainable food system adventures and find out about her upcoming foraging tours at ledameredith.com, as well as watch her foraging and cooking videos on YouTube.

INDEX

Italic page numbers indicate illustrations.